William Carmichael McIntosh

The Marine Invertebrates and Fishes of Saint Andrews

William Carmichael McIntosh

The Marine Invertebrates and Fishes of Saint Andrews

ISBN/EAN: 9783337416867

Printed in Europe, USA, Canada, Australia, Japan

Cover: Foto ©berggeist007 / pixelio.de

More available books at **www.hansebooks.com**

THE MARINE INVERTEBRATES

AND

FISHES

OF

ST. ANDREWS.

BY

W. C. M'INTOSH, M.D.,

F.R.S.E., F.L.S., C.M.Z.S., ETC.

EDINBURGH:
ADAM AND CHARLES BLACK.
LONDON:
TAYLOR AND FRANCIS, RED LION COURT, FLEET STREET
1875.

PRINTED BY TAYLOR AND FRANCIS,
RED LION COURT, FLEET STREET.

ALERE FLAMMAM.

TO

E. M.,

WHO SUPPLIED MOST

OF

THE RARE AND VALUABLE SPECIMENS,

This Work

IS DEDICATED

BY

HER SON.

PREFACE.

THE following pages may be considered a census of the collection of the Author, and therefore of little more than local value.

The synopsis of the Fauna is far from being complete, since circumstances have prevented much addition to observations made many years ago, when, perhaps, experience in such investigations was less matured, though opportunity might have been more ample. It may be regarded, indeed, as a list of the more conspicuous forms in each group; for with so wide a field nothing like an exhaustive catalogue could be attempted. The latter circumstance, however, may prove an encouragement to those who may be interested in the subject, from the facility with which additions can be made to the lists.

The beach at St. Andrews has afforded a harvest to numerous zoologists. The late Dr. John Reid, Professor of Medicine, made many original observations on the Nudibranchiate mollusks, molluscoids, and Polyzoa. The genial and distinguished Prof. Edward Forbes also knew the value of the specimens thrown ashore on the West Sands. The late Dr. G. E. Day (the accomplished successor of Dr. J. Reid), aided by Miss Otté, always encouraged his students to cultivate a knowledge of the riches of the bay. The late Dr. Fraser

Thomson, of Perth, Mr. E. Ray Lankester, the late Dr. Adamson, of St. Andrews, the late Mrs. Macdonald, Mr. Walker (the University Librarian), and many others have also specially interested themselves in the marine animals of St. Andrews.

My thanks are due to the following gentlemen for their valued aid in the several departments: viz., to Dr. Bowerbank in the Sponges; Mr. H. B. Brady, in the Foraminifera; Mr. Busk and the Rev. T. Hincks, in the Zoophytes and Polyzoa; the Rev. A. M. Norman, in the Echinoderms, and also, with Mr. Spence Bate and Mr. G. S. Brady, in the Crustacea; and Dr. Gwyn Jeffreys, in the Mollusks. I have also to remember the kind attention, in regard to the University Library and Museum, of Principal Tulloch, Professors Bell, Macdonald, and Heddle; while Mr. R. Walker, the Librarian, has supplied me with the names of a few additional fishes.

The Plates are the earlier efforts of my late sister; and the vignettes (with the exception of a sketch of the Harbour by my nephew) are reduced outlines of various pictures and coloured drawings by the same lady, who, with other relatives, was equally persevering in collecting and in portraying.

The lists for the most part appeared in the 'Annals and Magazine of Natural History' for the present year (February to December).

I have to acknowledge, in conclusion, the skilful touch of Mr. Ford in his rendering of the Plates. The Woodcuts were engraved in Edinburgh, by Mr. J. M. Corner, who spared no pains in their execution.

    Murthly,
    December 1874.

THE MARINE FAUNA

OF

ST. ANDREWS.

INTRODUCTION.

THE beach at St. Andrews combines smooth sandy flats with tide-worn ridges of rocks which freely communicate with the German Ocean; and the proximity to rich coralline ground renders the products of its storms peculiarly varied. An unbroken surface of pure sand extends from the estuary of the Tay past that of the Eden to the north-western border of the city. From this point the rocks run eastward in parallel rows—narrow sandy flats intervening between some of the ridges, which, with one exception, are all covered at high water. Lines of rocks having a similar arrangement fringe the Castle and Pier to the East Sands; then a coarse sandy and gravelly beach extends in a southerly direction about half a mile, after which the jagged rocky border passes round the eastern coast to the Frith of Forth.

The greater part of the sandy bay has a depth of less than 10 fathoms; for at this point the 20-fathom line bends outwards to the Bell rock. The whole region is thus comparatively shallow, and in contrast with that to the north of Arbroath Road, or with the Frith of Forth and the neighbouring coast on the south.

If the fine stretch of sand from the river Eden to the city (usually termed the West Sands in contradistinction to the East Sands which extend from the harbour southward) is only enlivened in summer by thousands of bleached heart urchins, broken shells, skeletons of plaice, frogfish, and haddock, or

B

in autumn by the jellies of the medusæ, the storms of winter
and spring wholly alter the aspect.   Immense banks of seaweeds mingled with black fragments of wood, coal, and
muddy matter cover the beach, which in many parts becomes
brilliantly phosphorescent at night from the zoophytes and
annelids on the blades of the tangles.   Amidst this débris are
vast numbers of sponges, zoophytes, shells, starfishes, annelids,
crabs, and fishes which have been swept from their various
habitats.   All storms are not equally prolific; they also vary
in regard to the abundance of the several groups—a feature
probably due to the direction of the wind and the invasion of
particular sites.   The waste of marine life in such storms does
not attract much notice; yet it is extraordinary and so constant
that it may be regarded to some extent as a check upon its
uninterrupted development.   It is, however, to be remembered
that even the autumnal ripple in the Outer Hebrides brings
countless swarms of *Salpæ, Velellæ, Medusæ,* and other forms
to die on the beach.

When the tide has receded after a severe storm the general
appearance (to select an example in October) is as follows :—
Besides the Fuci and other seaweeds, here and there banks
are formed almost entirely of tangles.   Now and then one is
attached to the shell of a living dog whelk, a cumbrous load
for so tiny a mollusk; others spring from polished pebbles
(bound together by their roots), from the shells of the great
whelk, *Tapes pullastra, Mactra solida,* horse mussels, and others,
or bear evidence of having been forcibly torn from the rocks.
The latter more readily happens when the surface of the rock
has been coated with *Balani, Lepraliæ,* and the purplish dif-
fused form of *Corallina officinalis,* to which the roots of the
tangles adhere—a fact which can be tested in a rock-pool.
In the interstices of the twisted roots many specimens of *Saxicava rugosa* occur, their habitations being more easily constructed, though less secure, than perforations in the rock
itself.   Porcelain crabs can scarcely be observed in greater
abundance, in company with the young of the shore and
spider crabs ; while a few sessile-eyed crustaceans try to escape
notice in the crevices tenanted by common and brittle stars.
The root-fibres also give shelter to specimens of dog whelks,

valves of the common mussel, *Venus striatula*, *Anomiæ*, and many fragments of shells and stones. Ascidians of various kinds are studded on the roots or in close array along the blades; and the stellate molluscoid forms (*Botryllus* and *Botrylloides*) are equally abundant. The coarse tunnels of *Thelepus* and *Terebella*, the finer tubes of *Sabellaria*, and the white coils of *Serpulæ* wind amongst the fasciculi; and Nereids and *Lepidonoti* creep into every protecting crevice. *Lepraliæ* and the ubiquitous *Membranipora pilosa* cover the dull surface of the tangles with silvery patches, intermingled with masses of *Alcyonium*, hydroid zoophytes, and shaggy fringes of seaweed. On the broad blades of *Laminaria saccharina* grow forests of *Obelia geniculata*, amongst which rare and curious nudibranchiate mollusks still find food and shelter after their perilous journey. The same seaweed is streaked with the tough gelatinous tubes of a little phosphorescent annelid (*Eusyllis*).

The chief representative of the sponges in these storms is *Halichondria panicea*, though occasionally *Chalina oculata* is equally abundant. The rugged masses of the former afford a favourite lurking-place for *Doris tuberculata*, porcelain crabs, and *Lepidonoti*; while in the general wreck it becomes hispid with the stiff sharp spines of the sea-mice, common and purple heart urchins. Sea-mats and corallines in general are plentiful. Numerous fragmentary *Medusæ* (chiefly *Aurelia aurita* and *Cyanea capillata*) strew the sands in company with *Pleurobrachia*, the parasitic *Hyperiæ* also being liberated from the former and swarming in the tide-pools. Anemones are common, either perched inside shells or mixed with the débris; and some of the rarer (such as *Peachia hastata* and the case-making *Edwardsia*) lie freely in the pools. *Alcyonium digitatum* occurs in great profusion, both loose and attached to various shells, or on the leathery tubes of *Sabella pavonia*; and the crevices of this zoophyte contain many specimens of *Tritonia plebeia*. Here and there are groups of large cockles (*Cardium echinatum*) exposing their pink feet through gaping valves; and in shallow pools the great whelk (*Buccinum undatum*) spreads out its speckled fleshy disk. Multitudes of other univalves and bivalves are scattered around, tufted with parasitic seaweeds and

corallines, or knobbed with *Balani* and masses of *Serpulæ.*
Swept from their haunts in the deep water and stranded
by the restless waves on the sandy beach, the shell-
fish—thrusting siphons or feet from their relaxing valves
in sickness, in search of safety or their native element—are
now pounced on by hooded and carrion-crows and gulls;
and the vacant and gaping valves frequently attest how
daintily they and the cunning oyster-catchers have fed on
the soft inhabitants. Hosts of young sea-urchins are found
on the beach alive; and the safety of the larger specimens
is attained by immersion in the soft débris. Occasionally a
bare portion of sand is speckled with numerous starfishes, from
the pale hues of the sand-stars to the brilliant scarlet of *Solaster
papposus* and the purples of *S. endeca* and *Astropecten irregularis.* Others slowly disentangle themselves from the wreck,
and, throwing out their sucker-feet, make active progress over
the tiny sand-ridges. *Ophiura lacertosa,* again, often buries
its disk from the gaze of the sea-fowl or for moisture, leaving
its arms projecting. Another echinoderm, *Cucumaria com-
munis,* is plentiful, though it can scarcely be distinguished
in the muddy débris. Vast multitudes of the sheaths of *Terebellæ* lie on the sands and especially collect on the bottom of
the Swilken burn; they are for the most part, but not always,
empty, the agitated waters having compelled these denizens
of the sand and gravel either to evacuate their homes in their
native sites or before reaching the beach; and hence the dis-
proportion between the tubes and their inhabitants. Thousands
of the annelid *Phyllodoce citrina* may be scooped from the
sand-pools. *Nereides* lurk in shells, sheaths, or anywhere for
shelter; sea-mice occasionally in myriads are indicated by their
splendid bristles amidst the débris; and in the pools lie the
curious *Ophelia limacina,* the golden-bristled spoon-worm, and
many other rare annelids. Sessile-eyed crustaceans also
swarm in the pools and tide-streams; while the larger forms
are represented by *Hyas,* the edible, shore, and hermit crabs.
Fishes are sometimes scarce—dead gurnards, plaice, and frogfish being the most conspicuous, though occasionally weevers,
sand-launces, viviparous blennies, gunnels, armed bull-heads,
suckers, father-lashers, and tadpole-hakes are also found.

The tidal rocks, again, which are mostly covered by sea-weeds, present a varied and prolific site for many species. The rock-pools are both frequent and picturesque; and they possess many undisturbed stones, often of large size, the under surfaces of which are most favourable for the growth and shelter of numerous forms—though of course they cannot be compared in this respect with the littoral stones at Herm, which have a profusion of rare crabs, annelids, ormers and other mollusks, polyzoa, hydroids, and the yellow, red, purple, green, and white sponges. The rocks may conveniently be divided into the East and West Rocks, the former stretching from the pier to the east, the latter from the pier westwards. The rock-pools in these resemble miniature oceans of surpassing beauty. The borders of the clear basins are luxuriantly fringed with different kinds of seaweeds—*Fucus serratus, Halidrys siliquosa, Furcellaria,* and *Chondrus,* and the broad blades of *Laminaria saccharina* hang sombrely downwards and trail on the bottom. These sober tints are here and there relieved by the fine green tufts of *Cladophora* and the broad *Ulva,* the dark red of *Ceramium rubrum,* and the pinkish hues of *Corallina.* The latter, in some of the clearest and finest pools, is not infested by the little white coils of *Spirorbis,* so that it is variegated only by its own whitish tips, which are in strong contrast in the quiet depths with the dull olive roots of the tangles.

The pools are tenanted by representatives of all the animal subkingdoms; and many species are in profusion. *Cotti,* gunnels, blennies, shannies, gobies, and sucking-fishes swim in the clear water or glide under the protecting seaweeds. Bunches of *Furcellaria* and *Fuci* afford a site for *Eolidæ;* and *Hyas, Carcinus,* and *Pagurus* nestle in their shade or perambulate the bottom. In some pools beyond the Maiden rock *Palæmon squilla* abounds, darting hither and thither with great velocity or poising its long antennæ over the seaweeds in graceful curves. Minute crustaceans and mollusks people the laminarian blades, while the roots give shelter to many of the forms previously mentioned. The larger mollusks (*Buccinum, Purpura,* and *Trochus*) crawl on the same leaves, or adhere in the society of limpets to the margins of stones. The under surface of the latter reveals a varied fauna. It is patched with *Cynthia grossularia* either singly or in groups, each with its brick-red tunic and prominent papillæ. A miniature forest of *Sertularia pumila* is carpeted by pinkish *Lepraliæ* and gelatinous molluscoids of diverse patterns and colours. White *Serpulæ* and the sandy tubes of *Sabellaria* are mingled with the coils of hosts of *Spirorbis*. Sponges of various species, developing *Medusæ*, molluscoids, nudibranchs, and other mollusks, starfishes and their ova, annelids and crustaceans, reward a careful search. Some shallow pools, again, are hollowed out of the surface of the bare rocks near high-water mark, *e. g.* near the Maiden rock ; and in each of these dwell a group of *Littorina rudis*, numerous sessile-eyed crustaceans, a solitary anemone (*A. mesembryanthemum*) and a few limpets, the latter bearing the most conspicuous vegetation on their shells.

Certain pools, especially near the pier, begin by a shallow margin, but end deeply against a perpendicular wall of rock, the bottom thus gradually shelving downwards. On the shallow border *Ceramium rubrum* and other seaweeds tuft the slopes, and are the favourite cover of many specimens of *Caprella tuberculata*, whose red bodies can scarcely be recognized amongst the branches. Each plant of *Ceramium rubrum*, indeed, is the abode of a marine colony. *Hippolytæ* cling to the boughs, their translucent bodies forming excellent subjects for the microscope, even where colour and outline are the only objects ; they exhibit lively motions of the antennæ, maxillipedes, and the ambulatory limbs ; while internally the flattened branchial and other whips vibrate with rapidity. *Idotea tricuspidata* darts from place to place, climbs up the branches, or remains stationary on a twig. Small sessile-eyed crustaceans alternately swim through the water, and run like monkeys along the boughs. Many little mussels with hairy valves hang motionless as fruit on the branches. *Rissoæ* move up and down the twigs—stretching their delicate white tentacula and blunt snouts, or vary their exercise by floating on the surface, shell downwards. Some detach themselves and pass swiftly to the side of the vessel and out of the water, thus forming a voluntary and sometimes useful segregation. *Lacunæ* also occur on the branches, and twist their muscular

feet and turn their shells with business-like pertinacity, mean-
while exposing their minute black eyes and feeling in all
directions. The little *Skenea planorbis* can just be distinguished with the naked eye as it crawls and twists on the
seaweed; it is less apt to leave the latter and its native element
than the *Rissoæ*. From the cells of *Membranipora pilosa* incrusting the branches of the seaweed, the bodies and tentacles
of the pretty polyps protrude. *Campanularia verticillata* and
C. volubilis expand their delicately barred arms over the
transparent cups—now contracting as prey is seized, or in-
stantly if touched, and again gently unfolding. *Hippothoa
divaricata*, which here (on seaweeds), according to Dr. Johnston, alters its shape, shoots its delicate white stems from a
branch of the *Ceramium*. Masses of the snowy sponge (*Leu-
cosolenia botryoides*) tuft the tips of others, while in the
vantage-ground afforded by a fork of the seaweed a colony of
Leptoclinum is seated. *Pedicellina echinata* here and there
attracts notice—some of the stems headless, as described by
Dr. J. Reid. When irritated the heads bend so as to touch
the basal portion of the stem or the surface of the seaweed.
In the deep part of the pool the tangles spring from the perpendicular rock and from stones at the bottom—waving their
rich fringes of phosphorescent *Obelia* on agitation of the water.
In the pool swim *Hippolyte varians* and *H. pusilla*, sessile-eyed
crustaceans, and *Mysidæ*; but the latter are not common in
pools north of the pier, apparently giving place in this
stormy region to *Hippolyte*. Besides the ubiquitous shore
crab, each pool is inhabited by a spined *Cottus*, whose fine
iridescent hues of silvery bluish green are displayed most
vividly in the water. The larval form called *Campontia
eruciformis* by Dr. Johnston swarms amongst the roots of
Corallina in summer, just as another insect-larva does amidst
the damp and decaying seaweeds on the sand at high-water
mark, and a third in the muddy fissures of the tidal rocks.

The soft sandstone and shale afford an ample field for the
perforations of *Pholas crispata*, *Saxicava rugosa*, and *Leucodore ciliata*. The fissures and chinks of the rocks, moreover, as on almost every part of the British shores, give
shelter to a large number, especially the annelids, which find

in the muddy or sandy crevices a safe retreat for their soft bodies, slender tubes, or muddy tunnels, and opportunities for capturing sufficient food at the free margin of the rock or from the ingoing currents.   It is chiefly in such localities that *Sipunculus Johnstoni* and swarms of *Leucodore* and *Nicomache* occur, while *Nereis cultrifera, Eulalia, Syllis,* and the nemerteans are also common.   Occasionally an *Idotea* is met with ; but by the general absence of the isopods these crevices are distinguished from those in the gneiss of the south and west, as in the Channel Islands and the Outer Hebrides; and they are especially distinguished from the former by the absence of *Pilumnus hirtellus, Arca,* the *Sabellidæ,* the *Eunicidæ* and their allies.   To these fissures certain boring annelids and *Saxicaea* chiefly retreat when the rocks do not afford a suitable medium for their perforations—though at St. Andrews there is free scope in this respect, from the sandstones and shales so soft as to be pitted deeply by the common limpet to those nearly as dense as granite.

If further, under the favourable ebb of a spring tide, the water has receded to an unusual degree, the observer may look down from the jutting rocks on the rich laminarian forests that flourish in a region of perpetual flood, and watch the silvery *Membraniporæ* and the amethystine streaks of *Heleion pellucidum* on their flattened bands, which wave and curl with every surge of the sea ; and just as the trunks of the forest-trees are incrusted with lichens, so feathery tufts of red algæ cluster on the stalks of the seaweeds.   Hosts of coal-fishes swim through and among them ; and entangled *Medusæ*, now at the mercy of the tide, are lifted on the rock and again washed off.   Scarlet Solasters and more soberly coloured Urasters contrast strongly with the dark water and olive-green seaweeds ; and nothing can be more beautiful than the light purplish pink of the sea-urchins as they progress by aid of their sucker-feet along rock, stone, or laminarian root.

Moreover, in the deeper water off the bay the fishermen occasionally secure rare fishes, entangle a shark, a bonito, a porpoise, a seal, or even a great northern diver in their nets, and bring many rare invertebrates to the harbour in their boats. The crab-pots, salmon-nets, and trawls further add their quota to swell the list of marine animals, and associate with the venerable city interests and ties of no common order even in this department alone.

The sea-margin at St. Andrews, like other parts of the east coast of Scotland, presents decided differences when contrasted with the northern, southern, and western shores, though many forms are common to all. Thus the laminarian zone at St. Andrews is much less luxuriant than that of the Zetlandic waters with the fine forests of gigantic tangles, amidst which there is a galaxy of animal life. The vegetation of its littoral zone is surpassed by the rich Fuci of the tidal rocks and the trailing masses of *Chorda filum* on the surface of the sea immediately beyond low-water mark in the Outer Hebrides. Its marine forms are placed under very different circumstances from those in the quiet voes of West Shetland, as at Cliff Sound and between the Burras, where the still sea-water is bridged by a single arch of a few feet. To represent the *Zostera*-fields of the west and south there are but a few *Confervæ*, *Ulvæ*, and *Porphyræ* attached to stones on the flat surfaces of the beach. There is no confusion of fresh and salt water as in the Hebrides, where from a hill-top the eye is quite unable to trace the intricate connexions of the endless lakes or distinguish the one element from the other at full tide—where, moreover, the breadth of the highway is the only separation in some cases between the rich vegetation of the fresh water, with its white and yellow water-lilies, and the swamps of *Zosteræ* and *Confervæ* of the salt. The calcareous rocks of the south, and the multitudes of worm-eaten boulders scattered on many parts of the shore, as in the Isle of Wight, form likewise a boldly marked contrast, which is heightened in some of the chalky bays by the semi-milky colour of the flowing tide (from calcareous admixture). Boring forms are very conspicuous in the latter rocks, but they are by no means confined to them; indeed they abound at St. Andrews. The muddy beach at the estuary of the Eden affords a site for the splendid mussel-beds; but (though *Corophium* is present in both) it cannot be compared with the tenacious greyish mud which sometimes, as at Herm, retains footprints so firmly that they are visible after

c

several tides, and swarms with *Sipunculus*, *Edwardsia*, and *Lumbriconereis*.

Within reach of the modern tide, also, it is interesting to find the remains of oceanic animals long since extinct—to see *Actinia mesembryanthemum* attached to a mass of encrinite stalks, *Littorina rudis* in groups on *Lingula*-shale, and the white coils of *Spirorbis* incrusting a nodule containing a fossil fish. Yet these features do not appear much out of place near a city whose pier is to a considerable extent constructed of the fine old stones and ancient oak which once formed part of the splendid pile of its cathedral.

On the whole the zoological features of St. Andrews are northern.

Subkingdom *PROTOZOA*.

Class RHIZOPODA.

Order FORAMINIFERA.

The beach at St. Andrews affords very ample opportunities for the study of the Rhizopods; but unfortunately, with the exception of the Spongiadæ, I am only able at present to give a list of a few common Foraminifera, which were kindly determined for me by Mr. H. B. Brady, of Newcastle, one of the highest authorities on the subject. They were obtained from shell-sand collected near the estuary of the Eden, and are forms common in shallow water and the littoral region.

Family **Miliolida**.

Biloculina depressa, D'Orb.
Triloculina oblonga, Mont.
Quinqueloculina seminulum, L.
—— *subrotunda*, Mont.
—— *secans*, D'Orb.

Family **Lagenida**.

Lagena sulcata, W. & J.
—— *globosa*, Mont.
—— *squamosa*, Mont.
Cristellaria crepidula, F. & M.
Polymorphina lactea, W. & J.
—— *gibba*, D'Orb.
—— *compressa*, D'Orb.

Family **Globigerinida**.

Textularia sagittula, Defr.
Truncatulina lobatula, Walker.
Rotalia Beccarii, L.

Family **Nummulinida**

Polystomella striatopunctata, F. & M.
Nonionina depressula, W. & J.

Mr. David Robertson, in the absence of Mr. Brady, kindly examined similar shell-sand, and also mud from the interstices of *Filigrana implexa* from deep water. To the foregoing list he adds *Discorbina globularis*, D'Orb., and gives the following species from the latter:—

Cornuspira foliacea, Phil.   Moderately common.
Quinqueloculina seminulum, L.   Rare.
—— *subrotunda*, Mont.   Rare.
Trochammina incerta, D'Orb.   Moderately common.
Polymorphina lactea, W. & J.   Rare.
Textularia sagittula, Defr.   Common.
Bulimina marginata, D'Orb.   Rare.
Discorbina globularis, D'Orb.   Moderately common.
Truncatulina lobatula, Walker.   Common.
Rotalia Beccarii, L.   Moderately rare.
Patellina corrugata, Will.   Rare.
Operculina ammonoides, Grou.   Rare.
Nonionina depressula, W. & J.

Order SPONGIADÆ.

The Sponges of St. Andrews are, perhaps, the least-investigated group, partly because a collection carefully made many years ago has been lost. In looking over those obtained since, Dr. Bowerbank has most kindly given his experienced aid in doubtful cases; and the description of the new species is solely his. The littoral forms are scattered in considerable profusion between tide-marks under ledges and stones, sometimes near high-water mark. Indeed, in the higher pools and tide-runs in the latter region they are often peculiarly luxuriant. The brightly coloured *Halisarca*, so abundant on the under surfaces of stones in the Hebrides, and the rarer botryoidal *Tethea* are unknown at St. Andrews, as are likewise the cup and turnip sponges of the Zetlandic seas. The greater luxuriance of the ubiquitous *Halichondria panicea* on the stems of the Laminariæ further characterizes the coast of the extreme west;

and the decay of the seaweed often leaves tubes of sponge from a foot to eighteen inches in length. In like manner the greatly increased size of *Grantia ciliata*, the vast abundance of *Hymeniacidon celata*, its beautiful arborescent patterns in the tide-worn shells, and its perforations in the limestone rocks are diagnostic of the warmer waters of the southern coast.

The classification of Dr. Bowerbank in his valuable work published by the Ray Society has been that followed in the list.

Suborder I. CALCAREA.

Grantia compressa, Fabr.; Bowerb. Brit. Sponges, vol. ii. p. 17.

Abundant on *Cynthia grossularia* under shelving rocks between tide-marks, and attached to the roots of Fuci and other seaweeds. It occasionally assumes an abnormal form, and has a broad attached surface under stones. Longest, 3 inches.

Grantia ciliata, Fabr.; Bowerb. vol. ii. p. 19.

Not unfrequent on laminarian roots cast on the West Sands after storms, and growing near low-water mark at the East Rocks. The species somewhat resembles a grain of oat removed from its husk.

Leucosolenia botryoides, Ellis & Sol.; Bowerb. vol. ii. p. 28.

Abundant on the under surfaces of stones in tidal pools, especially if large and little-disturbed. It frequently accompanies *Grantia compressa*.

Leuconia nivea, Grant; Bowerb. vol. ii. p. 36.

Found abundantly in the deeper tidal pools, under large

stones which have been long untouched. It covers spaces several inches square; and its margin is generally rounded and "finished" like the border of a lichen. Most of the specimens have their surfaces elevated into firm rugæ, resembling miniature mountain-ranges, some of the crests rising into flattened lobes ¾ inch in height. There are at least two varieties of this sponge—the first of which, besides the equiangular triradiate spicula of the skeleton, the minute acerate ones of the interstitial and dermal membranes, and the unicurvo-cruciform, has many spined acuate spicula of considerable dimensions and others of the same size approaching the fusiformi-spinulate character. In the other variety the latter kinds are so little developed, if present, as not to be distinguished from the ordinary minute acerate forms. In both, almost all the latter are distinctly spined.

Suborder II. SILICEA.

Hymeniacidon ficus, Esper; Bowerb. vol. ii. p. 206.

Occasionally from deep water, attached to dead shells. Clavate specimens frequently grow from the smaller end of *Dentalium entalis*. This species seems to frequent muddy ground.

Hymeniacidon celata, Grant; Bowerb. vol. ii. p. 212.

Abundant in shells from deep water, between the layers of which it tunnels its devious tracks. This is one of the main agents in causing the disintegration of dead shells.

Halichondria panicea, Pallas; Bowerb. vol. ii. p. 229.

Scarcely a stone can be lifted near low-water mark, amongst the rocks, but has a patch of this common sponge. Under the cavernous ledges overhanging rock-pools it spreads its structure over the dark red *Cynthia*, matting together seaweeds and corallines, and hanging in pendulous nodules on interwoven stalks of *Corallina officinalis* and Fuci. Near the Maiden Rock splendid specimens are found incrusting a square foot or two of rock in some of the quiet pools. It also abounds on the backs of crabs, such as *Hyas araneus* and *Inachus dorynchus*, covering the former so completely that it can scarcely be recognized except by its legs ; and besides the prominent oscula of the sponge, on this complex back are gaps for *Balani*, shells, and seaweeds. On the carapace of the latter species it forms a thinner coating, but is likewise grouped in little nodules on the legs. A mass as large as a good-sized apple surrounds the stem of *Chalina oculata*; and it is a common envelopment of various stones, mollusks, seaweeds, and tangle-roots. The usual colours of the sponge are yellow, brown, purple, green, and grey. In the interstices of the masses thrown on shore at the West Sands are to be found multitudes of marine animals, besides incorporated shells ; and the fine patches at the East Rocks are favourite feeding-grounds of *Doris tuberculata*. The forms of the spicula of this species are variable, some being much curved like a stretched bow, a few more or less inæquiacerate vermiculoid, besides, of course, the ordinary diagnostic spicula. The odour emitted on tearing it from the rock is characteristic, but causes no sneezing.

Halichondria, n. s.*

The following is Dr. Bowerbank's description :—" Sponge coating, thin. Surface smooth and even. Oscula more or less elevated, dispersed, margins thin. Pores inconspicuous. Dermal membrane aspiculous. Skeleton very irregular, rete mostly unispiculous, occasionally bi- or trispiculous ; spicula acerate, short and stout. Interstitial membranes aspiculous.

" Colour in the dried state light nut-brown. Examined in the living and dried states.

" The nearest alliance with the known species of the first section of our British *Halichondria* is with *H. regularis*. The spicula of the two species are as nearly alike in size and pro-

* Dr. Bowerbank has courteously named this species *H. M'Intoshii*, Bowerb.

portions as possible; but this is their only approximation to each other; in their other characters they differ to a considerable extent. The colour of *H. regularis* in the dried state is milk-white; that of *H. M'Intoshii* is nut-brown. Another important difference is, that while the skeleton of *H. regularis* is remarkable for its symmetry, that of *H. M'Intoshii* is irregular to a very considerable degree."

This form is not uncommon on the under surface of stones in tide-runs and somewhat muddy pools not far from high-water mark at the East Rocks. Its greyish brown colour, smooth surface, and prominent, well-defined oscula distinguish it at first sight from its allies.

Halichondria incrustans, Esper; Bowerb. vol. ii. p. 249.

Occasionally found under stones near low-water mark, especially at the East Rocks. It forms a thickish crust; and the spicula very much resemble one of the knobbed walking-sticks which taper from above downwards.

Suborder III. KERATOSA.

Chalina oculata, Pallas; Bowerb. vol. ii. p. 361.

Thrown in great profusion on the West Sands after storms; and small specimens are also found under the ledges of rocks near low-water mark. The shape of the specimens varies much: some are flattened and much divided into branches of various sizes, either narrow or broad; others have their branches matted together so as to form a connected and somewhat coarse "gorgonian" appearance, more or less separated at the tip. In some the branches arise mostly from one side of an unbroken prolongation of the sponge-tissue. One grows on a valve of *Mytilus modiolus*, and has a mass of *Halichondria panicea* round a branch at its base. Another envelops the stem and branches of *Delesseria sanguinea*, the leaves of which appear here and there from the centre of the sponge. Many are attached to small rolled stones.

Those from the beach are loaded with sand, spines of the common and purple heart urchins, bristles of the sea-mouse; and many starfishes seek refuge in their interstices.

Chalina limbata, Bowerb.; vol. ii. p. 373.

Not uncommon under stones in tidal pools, either coating the surface of the stone or attached to the stems of *Corallina officinalis*.

R.M. f^ANHL.   J.M.L.

D

Subkingdom *CŒLENTERATA.*

Class HYDROZOA.

The Hydroid Zoophytes of St. Andrews are chiefly procured from the deep water of the bay, though a few appear
between tide-marks. Many are found in great profusion.
Compared with the southern shores, as at Devon and Corn-
wall *, the majority of the Hydroids are equally common in
both localities, some occur more frequently in the one than
in the other, while a third series is more characteristic of
each area. Thus *Sertularella rugosa, Sertularia cupressina,
Thuiaria thuja,* and *Halecium muricatum* appear to be more
abundant at St. Andrews than in the south; on the other
hand, *Sertularia argentea* and *Obelia dichotoma* are probably
more plentiful in the latter, together with the appearance of
Tubularia at the extreme margin of low water. The characteristic forms in the south are *Corymorpha nutans, Aglao-
phenia pluma, A. pennatula, Ophiodes mirabilis, Diphasia
pinnata,* and an abundance of the species of *Plumularia.* At
St. Andrews *Sertularia filicula, S. fusca, Tubularia coronata,
Cuspidella humilis,* and *Halecium labrosum* afford distin-
guishing features. Moreover, instead of the tufted *Clava
squamata,* so common on the littoral Fuci of the western coast,
we have *C. multicornis* at St. Andrews on the under surface
of stones; the splendid *Corymorpha nutans* of the sandy voes,
and the rich tufts of littoral Corynidæ and *Gonothyræa* of the
Zetlandic region, are likewise wholly absent. Amongst the
Hydromedusæ, *Sarsia prolifera,* Forbes, occurs occasionally;
and *Thaumantias pilosella,* Forbes, in great abundance on the
surface of the bay in autumn.

The habit of the zoophytes affords many interesting facts,
especially in regard to the profusion of parasitic structures.
The roots of the polyparies spring from diverse shells, stones,

* J. & R. Q. Couch, in their 'Cornish Fauna;' the elaborate catalogue
of the Rev. T. Hincks in the 'Ann. & Mag. Nat. Hist.' 1861-62; and
Mr. Parfitt's Devonshire Catalogue published in 1866.

crabs, submerged sticks and branches. One of the most curious examples found by the fishermen in the bay consisted of a stout branch of a thorn-tree, about four feet in height, which had large specimens of *Balanus Hameri* and Ascidians clustered like living fruit on the main trunk and branches, and lobulated and club-shaped masses of *Alcyonium* covering the more slender twigs and overrunning the neighbouring Cirripedes; while *Obelia* fringed most of the branches, here and there giving place to the shorter coating of *Sertularia*, stunted *Tubularia*, or the downy *Clytia*. Hosts of other animals occurred on the congenial site—tubes of *Thelepus* and *Serpula*, *Anomia*, *Saxicava*, *Xylophaga*, *Lepralia*, *Cellepora*, and *Tubulipora* representing the sedentary forms, sessile-eyed Crustaceans and Starfishes the free. Indeed the production formed a compendium of marine zoology that took much time and trouble to investigate. The rapidity of growth of the larger specimens (the *Balani* being as large as walnuts) was shown by the condition of the wood and bark, and the presence of many delicate twigs. This is also seen in the case of slender branches of the common currant-bushes, which are brought to land in good preservation yet densely fringed with *Obelia longissima* and studded with large ascidians. The zoophytes themselves are subject to many parasitic inroads from sponges, Foraminifera, other zoophytes, various Polyzoa, Ascidians, Nudibranchs and their ova, young mussels and *Anomiæ*, the ova of *Pycnogonum*, Annelids and their tubes (hyaline, gritty, and calcareous), and minute Cirripedes.

In the following list the arrangement and nomenclature of the Rev. T. Hincks in his recent beautiful work on the Hydroida is adopted.

Order I. HYDROIDA.

Suborder I. **ATHECATA.**

Fam. 1. **Clavidæ.**

Genus CLAVA, Gmelin.

Clava multicornis, Forskål; Hincks, Brit. Hydroid Zoophytes, vol. i. p. 2.

Frequent under stones in pools near low-water mark, and

growing on *Cynthia grossularia* under the cavern-roofs ; but
it is not seen on the littoral seaweeds, as is *Clava squamata*
on the shores of the Hebrides and the western and other coasts
of Scotland. The tentacles show a slightly enlarged sucker-
tip.

Fam. 2. Hydractiniidæ.

Genus HYDRACTINIA, Van Beneden.

Hydractinia echinata, Fleming ; Hincks, Brit. II. Z.
vol. i. p. 23.

Abundant on *Fusus islandicus*, *Natica*, and other univalve
shells cast on shore after storms. The outer lip in the shells
inhabited by hermit crabs is frequently prolonged into a horny
membrane, as mentioned by Dr. Johnston.

Fam. 9. Eudendriidæ.

Genus EUDENDRIUM, Ehrenberg.

Eudendrium rameum, Pallas ; Hincks, Brit. II. Z.
vol. i. p. 80.

Plentiful in the deep water of the bay, attached to shells and
masses of *Balani* and *Serpulæ*. A fine specimen measured
9 inches high ; and the breadth of the branched portion was
8 inches.

Eudendrium capillare, Alder ; Hincks, Brit. II. Z.
vol. i. p. 84.

Fine tufts are occasionally found on the stems of *Antennularia ramosa*, interwoven with other zoophytes, from deep
water. The specimens had no short branches ; all were much
elongated, and the polyps terminal. Some slight rings exist
here and there on the main stems at the base ; those at the
origin of each branch are very distinct.

Fam. 11. Tubulariidæ.

Genus TUBULARIA, Linnæus.

Tubularia indivisa, L. ; Hincks, Brit. II. Z.
vol. i. p. 115.

Common in deep water. One of the large specimens

springs from an agglutinated basis of the valves of *Pecten opercularis* and gravel, 8 inches in diameter; and the gigantic tuft had tubes 11 inches in height. It also sometimes fixes the valves of a living *Mytilus modiolus* so as almost to prevent motion.

Tubularia coronata, Abildgaard; Hincks, Brit. H. Z. vol. i. p. 119.

Abundant in deep water. I am obliged to Prof. Allman for discriminating wrinkled specimens of this species, in 1863.

Suborder II. THECAPHORA.

Fam. 1. Campanulariidæ.

Genus CLYTIA, Lamouroux.

Clytia Johnstoni, Alder; Hincks, Brit. H. Z. vol. i. p. 143.

Abundant on *Alcyonidium hirsutum* and seaweeds in the pools near low-water mark, as well as coating the stems of *Laminariæ* with a hairy fringe fully half an inch in height. In a fine example of the latter many of the stems possess one or two branches, and the gonothecæ here and there have a stalk composed of several rings.

Genus OBELIA, Péron & Lesueur.

Obelia geniculata, L.; Hincks, Brit. H. Z. vol. i. p. 149.

Common on laminarian blades thrown on the West Sands after storms, forming a miniature cover amidst which many Nudibranchs find food and shelter. It occurs plentifully also on *Halidrys siliquosa* and other seaweeds near low-water mark, and on crabs. In the interior of many of the gonothecæ are the young of a Pycnogonidian.

Obelia longissima, Pallas; Hincks, Brit. H. Z. vol. i. p. 154.

Abundant in deep water. It bristles on every branch or fragment of wood which has been submerged for some weeks. It appears also in a very interesting condition in the peculiar rounded balls formed by the rolling action of the waves on the beach; these zoophytic masses are either spherical or rounded-oblong, and the fibres are firmly felted together *. In this state the present species is stripped of its minute branches, and feels bristly and crisp. The same rolled masses (also chiefly composed of an *Obelia* allied to the present form) were brought from the shore of a New-Zealand bay by Dr. Lauder Lindsay, who kindly sent them to me. They are formed in a similar manner to the well-known balls in Loch Tay, where the rolling action of the waves produces perfectly round masses, often as large as a spherical shot of thirty or forty pounds, composed of the linear leaves of the larches and pines which shade its margin. Miss M'Leod, of Paible, brought me spherical masses of a similar description from a freshwater lake in South Uist, the species in this case, according to Prof. Dickie, being *Cladophora glomerata. O. longissima* affords a favourite site for young mussels.

Obelia dichotoma, L.; Hincks, Brit. H. Z. vol. i. p. 156.

Not common; parasitical upon a piece of seaweed from the laminarian region, and reaching about 3 inches high.

Genus CAMPANULARIA, Lamarck.

Campanularia volubilis, L.; Hincks, Brit. H. Z. vol. i.
p. 160.

Common on crabs, the stems of *Sertularia argentea* and other zoophytes from deep water. It is a smaller and more delicate species than *C. verticillata*. The shape of the cup and the very distinct " spherical ring " below distinguish it when the gonothecæ are absent.

Campanularia Hincksii, Alder; Hincks, Brit. H. Z.
vol. i. p. 162.

Occasionally found on the stems of *Antennularia antennina* from the deep water of the bay. This species presents certain

---

* One of these masses so closely resembled the chignon lately in vogue that it was secretly used by a patient for this purpose; and I learned that it was only the disagreeable abundance of sand in its tissue that saved it from further duty in this respect.

variations. In some the stem is nearly smooth from the base to the cup, where there are only a few slight twists; in others there are several distinct though irregular rings or twists at the base, a few about the middle of the stem, and others at the base of the calycle; in almost all there is one very distinct ring at the base of the latter, as Mr. Alder shows in his figure *. There is also a peculiar hollow at the base of the calycle; but this cannot be called a ring.

Campanularia Hincksii.

Campanularia verticillata, L.; Hincks, Brit. H. Z. vol. i.
p. 167.

Common in deep water. Many specimens were also found in the stomach of *Echinus esculentus* from the laminarian zone.

Campanularia flexuosa, Hincks, Brit. H. Z. vol. i.
p. 168.

Not uncommon on the under surfaces of stones near low-water mark. The peculiar zigzag form of the stem, with the arms of the forks tending in opposite directions, together with the short, broad, and smooth-edged hydrothecæ, are character-istic. The long pedicels of the hydrothecæ had their central smooth portions peculiarly flattened out, so as almost to assume a fusiform aspect.

Campanularia raridentata, Alder; Hincks, Brit. H. Z.
vol. i. p. 176.

Occasionally found on *Antennularia antennina* and other zoophytes from the coralline ground. The form agrees in most respects with the published description. The calycle is very narrow and deep, with six to eight large teeth on the margin; stalk rather slender (much more so than in *Clytia Johnstoni*), with several distinct rings below the cup, and many less distinct towards the base. The peculiar slenderness of the stalk, the

* Catalogue of the Zoophytes of Northumberland and Durham, pl. ii. fig. 9.

length of the cup, and the small number of teeth are the characteristic features. Specimens which resemble *Clytia Johnstoni* occasionally grow in proximity; and some intermediate forms occur.

Genus GONOTHYRÆA, Allman.

Gonothyræa Lovéni, Allman; Hincks, Brit. H. Z. vol. i. p. 181.

Abundant on *Sertularia abietina* and *Diphasia rosacea* from deep water. The exceeding delicacy of the free margins of the hydrothecæ, even in good spirit-preparations, renders it difficult to say whether they are (or were) notched or smooth. It was only by a comparison of observations on many examples that the peculiar crenations were understood, as none showed more than a few, and the majority none at all. The appearance of the gonothecæ, however, is characteristic.

Gonothyræa gracilis, Sars; Hincks, Brit. H. Z. vol. i. p. 183.

Plentiful on *Tubularia indivisa*, from deep water, amongst *Clytia Johnstoni*, on the tests of *Ascidia sordida*, on *Scalpellum vulgare*, *Stenorhynchus rostratus*, and *Cellepora pumicosa*. The capsules are large, translucent, and borne on a ringed stalk. Growing as this did amongst *C. Johnstoni*, it at first seemed to be a branched variety of the latter; but the peculiar nature of the branching and the structure of the gonothecæ, which were chiefly borne on the stems, distinguished it on closer scrutiny. Moreover the hydrothecæ of this species, contrasted with *C. Johnstoni*, are much larger and deeper.

Fam. 2. Campanulinidæ.

Genus OPERCULARELLA, Hincks.

Opercularella lacerata, Johnston; Hincks, Brit. H. Z. vol. i. p. 194.

Abundant on the stems of *Plumularia pinnata*, *Obelia longissima*, and other zoophytes, and amongst *Clytia Johnstoni* on the stems of *Laminariæ*; Prof. John Reid also found it on *Scrupocellaria scruposa*. This species presents two well-marked varieties, which occur together on the same stem :—

(a) hydrothecæ on simple stalks of variable length, viz. from three to nine rings; and (b) branched stems of some height, with the alternate stalks of the hydrothecæ composed of from six to more than a dozen rings. Moreover, in these branched forms it is not uncommon to see more than one pedicel arise at the same fork, so as to cause the observer to fancy he is viewing the *Campanulina turrita* of Prof. Wyville Thomson; only the hydrothecæ are much shorter in proportion to the length of the teeth. Some examples on a laminarian stalk had very long stems. The hydrothecæ in all very closely resembled those on Dr. Allman's *Campanulina repens* (Hincks, Brit. H. Z. vol. i. p. 189, pl. xxxviii. fig. 1). No gonothecæ were observed.

Fam. 4. Lafoëidæ.

Genus LAFOËA, Lamouroux.

Lafoëa dumosa, Fleming; Hincks, Brit. H. Z. vol. i. p. 200.

Common on various zoophytes from deep water. Some varieties of this species have short stalks of one or two whorls supporting the hydrothecæ; but they are not quite so long as those described under *L. fruticosa*, Sars, and the intermediate forms show that they are to be referred to the present species.

Lafoëa fruticosa, Sars; Hincks, Brit. H. Z. vol. i. p. 203.

Occasionally on zoophytes from deep water, especially *Sertularia filicula*. The pedicels of the hydrothecæ have from three to five rings.

Besides the above there are several microscopic forms closely allied, which creep along the stems of various zoophytes. A sessile form is common on *Crisia eburnea*, and a stalked species on *Scrupocellaria scruposa*.

Genus CALYCELLA, Hincks.

Calycella syringa, L.; Hincks, Brit. H. Z. vol. i. p. 206.

Abundant on the stems of *Hydrallmania falcata* and other zoophytes from deep water.

E

Genus CUSPIDELLA, Hincks.

Cuspidella humilis, Hincks, Brit. H. Z. vol. i. p. 209.

Not uncommon on the tests of *Ascidia sordida*, and on the valves of *Psammobia* and other shells, from deep water. The tests of Ascidians are the seat of a reticulated growth with numerous minute club-shaped processes rising from the creeping stem which is associated with *C. humilis*.

Genus FILELLUM, Hincks.

Filellum serpens, Hassall; Hincks, Brit. H. Z. vol. i. p. 214.

Abundant on the stems of *Sertularia abietina* and *H. falcata* from deep water.

Fam. 6. Coppiniidæ.

Genus COPPINIA, Hassall.

Coppinia arcta, Dalyell; Hincks, Brit. H. Z. vol. i. p. 218.

Common on the stems and branches of *Sertularia abietina* and *Hydrallmania falcata*.

Fam. 7. Haleciidæ.

Genus HALECIUM, Oken.

Halecium halecinum, L.; Hincks, Brit. H. Z. vol. i. p. 221.

Plentiful in deep water, though somewhat less common than the next species. Young specimens under an inch in height sometimes occur, which in spirit quite agree with the Rev. A. M. Norman's description of *H. sessile* (Hincks, *l. c.* p. 229, pl. xliv. fig. 2), with, of course, the exception of the polyps. In these cases the hydrothecæ do not seem to be fully developed; but they show the row of dots below the margin. Specimens are also seen in which one or two of the hydrothecæ are better developed at the base of the stem, while all the rest are in the condition described by Mr. Norman. It would appear to be doubtful if the mere elongation of the polyps would constitute specific distinction, any more than the fact that the branches are not in the same plane. Some are slightly ringed.

Halecium muricatum, Ellis and Solander; Hincks, Brit. H. Z. vol. i. p. 223.

This is the common *Halecium* from the deep water of the

bay. Most of the specimens show a ring or two on the stem above the calycles.

Halecium Beanii, Johnston ; Hincks, Brit. H. Z. vol. i. p. 224.

Not uncommon in the coralline region, attached to other zoophytes and to the tests of Ascidians. Young examples have piles of little cups on the hydrothecæ like those on the beautiful southern *H. tenellum,* Hincks.

Halecium labrosum, Alder ; Hincks, Brit. H. Z. vol. i. p. 225.

From the deep-sea lines of the fishermen. Rather rare.

Fam. 8. Sertulariidæ.

Genus SERTULARELLA, Gray.

Sertularella polyzonias, L. ; Hincks, Brit. H. Z. vol. i. p. 235.

Abundant in the deep water of the bay, generally attached to the roots of other corallines. All the specimens seem true to the description of the species; at least there is no tendency to wrinkles in the hydrothecæ so far as examined.

Sertularella rugosa, L. ; Hincks, Brit. H. Z. vol. i. p. 241.

Common under stones near low-water mark, and thence to deep water, on seaweeds, *Flustra foliacea,* &c.

Sertularella tenella, Alder ; Hincks, Brit. H. Z. vol. i. p. 242.

In profusion on *Sertularia abietina* and other zoophytes from the coralline ground.

Genus DIPHASIA, Agassiz.

Diphasia rosacea, L. ; Hincks, Brit. H. Z. vol. i. p. 245.

Abundant in deep water on other corallines, such as *Hydrallmania falcata, Thuiaria thuja, Halecium,* &c. The specimens are large and luxuriant. Mr. Hincks is right in stating that the male gonothecæ have eight longitudinal ridges, and not six as Dr. Allman says.

Diphasia tamarisca, L. ; Hincks, Brit. H. Z. vol. i. p. 254.

Not uncommon in deep water attached to shells and stones.

E 2

Genus SERTULARIA, Linn.

Sertularia pumila, L.; Hincks, Brit. II. Z. vol. i. p. 260.

This is the common Sertularian under stones in rock-pools at and near low-water mark at St. Andrews; but it is less luxuriant than on the Fuci of the western coasts. It forms a miniature forest on the under surface of the stones in quiet places, and is a favourite haunt of *Eolis viridis* and other Nudibranchs. It presents the peculiarity in such situations that the stem is not always contracted above the hydrothecæ, but not unfrequently these follow each other without the constriction. In some preserved specimens the hydrothecæ contained a number of large nucleated cells, having apparently a thickened and regularly crenated cell-wall; these cells varied in size; and some also occurred in the centre of the stem.

Sertularia operculata, L.; Hincks, Brit. II. Z. vol. i. p. 263.

Not uncommon; on seaweeds at and beyond low-water mark, but chiefly procured on the West Sands after storms. Its comparative scarcity is in marked contrast with its profusion on our western coasts, where almost every laminarian root and stalk are clothed with dense tufts.

Sertularia filicula, Ellis and Solander; Hincks, Brit. II. Z. vol. i. p. 264.

This, perhaps, is the most abundant Sertularian next to *S. abietina* from deep water. Dried specimens, when carefully laid out, show a somewhat rectangular arrangement of their terminal branches. Good examples have also been procured from the stomach of the cod.

Sertularia abietina, L.; Hincks, Brit. H. Z. vol. i. p. 266.

Very common, and occasionally reaching the height of 9 inches; fine tufts occur on *Mytilus modiolus*. This species is a favourite seat of many parasites, such as other hydroid zoophytes, calcareous corallines, *Spirorbis*, *Alcyonidium*, *Coppinia*, &c. From its attachment to living mollusca (*Anomia* and others) it is not unfrequently swallowed by the cod.

Sertularia argentea, Ellis and Solander; Hincks, Brit. H. Z. vol. i. p. 268.

Not common; from the deep-sea lines of the fishermen. This seems to be a form more characteristic of our southern shores.

Sertularia cupressina, L.; Hincks, Brit. H. Z. vol. i. p. 270.

Very plentiful in the coralline region—sometimes reaching the length of 18 inches. Besides the ordinary form there are two branched varieties. In the first, numerous secondary polyparies spring from the ordinary dichotomous branches, each twig so burdened being very little thicker than the ordinary forms, and bearing in the usual manner for some distance the hydrothecæ, which gradually become obsolete; this secondary trunk assumes considerable dimensions, with jointed stem and dichotomous branches, like an independent specimen. In the other variety the main stem itself splits into two divisions, or the secondary trunks throughout are directly connected therewith.

Sertularia fusca, Johnst.; Hincks, Brit. H. Z. vol. i. p. 272.

A single fine specimen only has yet been procured, in the deep water of the bay. Mr. Alder correctly observes that this form leads us to *Thuiaria*.

Genus HYDRALLMANIA, Hincks.

Hydrallmania falcata, L.; Hincks, Brit. H. Z. vol. i. p. 273.

One of the most abundant hydroid zoophytes from the coralline ground. Its form varies from the elongated spiral to the broadly branched condition; and it is frequently loaded with parasitic zoophytes, both horny and calcareous. It is a favourite site for Nudibranchiate Mollusca and their ova; and minute Annelids construct their tubes on every convenient bough. Young specimens are plentiful also under stones between tide-marks, where their habit differs considerably from the foregoing, having the form of a simple straight pinna, generally coated with parasitic structures, both animal and vegetable.

Genus THUIARIA, Fleming.

Thuiaria thuja, L.; Hincks, Brit. II. Z. vol. i. p. 275.

Common; chiefly frequenting dead valves of *Cyprina,* *Pecten,* and *Tapes,* as well as stones—shooting its long stems upwards (occasionally to the length of 14 inches) amidst masses of the tubes of *Serpula, Thelepus,* and other Annelids, and patches of *Alcyonium.* In some examples a short secondary stem branches from the main trunk near the base. Parasitic upon the stems are numerous other corallines, such as *Diphasia rosacea,* which clothes anew the bare zigzag trunk with a more silky fringe than nature originally provided; rough crusts of *Cellepora* or the spreading *Alcyonium* and *Alcyonidium* entirely surround it; while occasionally a long tunnel of *Thelepus* is glued from the base to the branching portion. Now and then it occurs in the stomach of the cod.

Fam. 9. Plumulariidæ.

Genus ANTENNULARIA, Lamarck.

Antennularia antennina, L.; Hincks, Brit. H. Z. vol. i. p. 280.

From the deep water of the bay; common, but less so than the next species. Fine tufts reach a height of fully 11 inches. In a curious example a number of simple straight stems proceed from the upper edge of a fragment of an old trunk.

Antennularia ramosa, Lamarck; Hincks, Brit. II. Z. vol. i. p. 282.

Common in deep water, whence it is usually brought by the fishermen's lines.

Genus PLUMULARIA, Lamarck.

Plumularia pinnata, L.; Hincks, Brit. II. Z. vol. i. p. 295.

Frequent in deep water, and often reaching the height of 7 inches. A tall variety is found in which no spines are present on the gonothecæ. It sometimes occurs on *Stenorhynchus rostratus* between tide-marks.

Plumularia Catharina, Johnst.; Hincks, Brit. H. Z. vol. i. p. 299.

Common on Ascidians, tubes of *Thelepus*, and the roots of other corallines in deep water.

Plumularia frutescens, Ellis & Solander; Hincks, Brit. H. Z. vol. i. p. 307.

Occasionally thrown on the West Sands after storms, and also brought in by the deep-sea lines of the fishermen. The smaller specimens are pale. One example is 6 inches in height, and broadly branched.

Order MEDUSIDÆ.

The Medusidæ abound chiefly in autumn in the bay, the most conspicuous amongst the larger forms being *Aurelia* and *Cyanea*, the former often occurring in such numbers as to form a closely packed layer on the surface of the sea over considerable areas; and though not in the dense party-coloured masses of various species occasionally seen in the Hebrides, still they form an interesting feature. At certain points the bay in quiet weather is quite purplish with thousands, many of which are loaded with ova; and through the transparent umbrellas the abdominal feet of the parasitic Hyperidæ are observed in constant vibration. Occasionally, whether from accident or design, one specimen is found adhering to the umbrella of another, and is thus carried through the water. Moreover, on many of the stones at the East and West Rocks, near low-water mark, a "*Hydra tuba*" is found, which may be the hydroid condition of the foregoing. This pretty little white structure, developed from the ova of *Aurelia* and its allies, can be observed in all stages not only throwing out lateral buds like a *Hydra*, but by transverse fission dividing into a series of saucer-shaped bodies which ultimately assume the form of the adult *Aurelia.* This form, it is well known, formed the subject of valuable observations by the late Prof. M. Sars, and afterwards, amongst others, by the late Dr. John Reid, who obtained his examples at St. Andrews.

Hydra tuba.

On the whole we lack at St. Andrews the splendid profusion of the swimming jellies occasionally met with on our western shores, and especially in the Outer Hebrides, to which a favouring wind and tide sweep them from the warmer area of the Gulf-stream beyond, in company with *Ianthina* and the Pteropods. Amongst these the strange and beautifully tinted *Diphyes* is seen darting hither and thither amongst the brilliant blues of its brethren with its trailing fringes of bright orange polypites; and on the lonely western shores, as at Monach, countless myriads of the little *Velella* are tossed in autumn on the sand.

Mr. Darwin *, referring to the colours of certain Invertebrate animals, thinks that it is doubtful if such serve as a protection; but he goes on to observe that the perfect transparency of the Medusæ, " many floating mollusca, crustacea, and even small oceanic fishes partake of this same glass-like structure," and that " we can hardly doubt that they thus escape the notice of pelagic birds and other enemies." It seems to me somewhat

* Descent of Man, &c. vol. i. p. 323.

difficult to say what will escape the eye of a pelagic bird, such
as gull, guillemot, or hawk-like tern. Their keen eyes
distinguish very indistinct objects—for instance, the nucleus
of *Salpo runcinata*, and the minute and almost transparent
bodies of the young fishes that flit amongst the splendid masses
of swimming jellies (Molluscan and Cœlenterate) which sometimes throng our western shores. The mere tremor of the
water is almost sufficient to attract such acute and skilful
marauders. Moreover the statement of the great naturalist
is incomplete without the appendix that many of the Medusæ
and Hydromedusæ are brilliantly coloured and, in addition,
phosphorescent, the latter property likewise characterizing the
translucent *Pyrosoma*, and that my distinguished friend Prof.
Wyville Thomson regards the luminosity of marine animals as
a provision of nature for attracting their enemies in the abysses
of the ocean, or for throwing a flood of light on their own prey.
I have already* shown my reasons for believing that the theory
of the latter author is open to doubt, and shall make a few further
remarks on the subject under the Annelida. If the notion had
been promulgated that the sexes in the abysses of the ocean used
their light to attract each other, and thus had a better chance of
continuing the race, perhaps more might have been said in its
favour.

Genus AURELIA, Pér. & Les.

Aurelia aurita, O. Fabr.

Abundant in autumn and often so late as November.

Genus CYANEA, Pér. & Les.

Cyanea capillata, Eschsch.

Common in autumn.

Order LUCERNARIIDÆ.

(*Calycozoa*, R. Leuck.).

Genus LUCERNARIA, O. F. Müller.

Lucernaria auricula, O. Fabr.

[Plate III. figs. 11 & 12.]

Frequent on Fuci near the commencement of the East
Rocks, and occasionally at the West Rocks. It is as common
as in the south.

* Ann. & Mag. Nat. Hist. 1872, ser. 4, vol. ix. p. 2.

F

Class CTENOPHORÆ.

Two representatives only are found at St. Andrews, viz. a species of Beroid and a *Pleurobrachia*. The former occurs in vast swarms in July, indeed almost as plentifully as in the Zetlandic seas, and is easily procured by the hand-net from the rocks or at sea. The latter is equally abundant from August to the end of autumn, and even in winter, occasionally filling the towing-net or the dredge in the bay, and thrown ashore after storms on the West Sands. Few objects are more engaging than one of these spherical jellies in a clear glass vessel of sufficient size to exhibit the matchless mechanism of of its complex tentacles and the splendid iridescence of its locomotive rows.

Order SACCATÆ, Agassiz.
Genus PLEUROBRACHIA, Flem.
Pleurobrachia pileus, Eschsch.

Abundant. It eagerly devours *Carcinus mœnas* in the zoëa-stage.

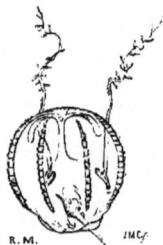

Pleurobrachia pileus, swimming downwards after engulfing a zoëa.

Order BEROIDÆ, Ggbr.
Genus IDYIA, Fréminville.
Idyia cucumis, O. Fabr.

Occasionally in large numbers in July and August.

Class ACTINOZOA.

Though the total number of species of this class at St. Andrews is small, many occur in great abundance, and especially such cosmopolitan forms as *Actinia mesembryanthemum*, *Tealia crassicornis*, and *Actinoloba dianthus*. The frequent occurrence of *Sagartia troglodytes*, again, at St. Andrews, distinguishes it from the shores of the extreme south, as at Guernsey. We have not, moreover, the fine *Anthea cereus* of the west and south, which, for instance, in the quiet creeks of the Outer Hebrides studs the stems and blades of the tangles at the border of the littoral zone, the beautiful greenish purple tentacles gently waving with every swell of the tide; neither is the gaudily tinted *Sagartia parasitica*, so characteristic of some of our southern shores, to be found between tidemarks, nor *Adamsia palliata* in deep water. *Corynactis*, the stony corals, *Zoanthus*, and the northern free-swimming *Arachnactis albida* are entirely absent. The places of these are filled by swarms of the common forms above mentioned, and by some of the rarer types, e. g. *Edwardsia*, *Cerianthus*, and *Peachia*, which seem to be characteristic of sandy beaches. A remarkable example * of the latter turned inside out occurs in my collection. It was mistaken for a curious polyp with beautifully arranged longitudinal and transverse muscular bundles, and was found inserted in a tunnel in the sand in this condition in Cobo Bay, Guernsey. It is simply a large *Peachia* everted.

Amongst the Alcyonarians the phosphorescent *Pennatula* occurs in great beauty, and replaces the *Pavonaria* of the west, while with *Virgularia* it also affords a diagnostic mark from the south. The fine Gorgoniadæ of the latter region, again, have no representatives at St. Andrews.

* I am indebted to Dr. Cooper, of St. Peter Port, for the specimen.

Order I. ZOANTHARIA.

Suborder ACTINARIA.

Fam. 2. Sagartiadæ, Gosse.

Genus 1. ACTINOLOBA, Blainville.

Actinoloba dianthus, Ellis; Gosse, Brit. Anem. p. 12,
pl. 1. fig. 1.

[Plate I.]

Common in the débris of the fishing-boats, and thrown
ashore after storms attached to sticks and shells. Young specimens (Plate II. fig. 8, and woodcut, p. 39) occasionally
appear on stones at extreme low water, and when very
hungry greedily swallow green seaweeds. Some expand
the disk like a *Doris*. or *Lamellaria*, and float on the surface
of the water.

Genus 2. SAGARTIA.

Sagartia troglodytes, Johnston; Gosse, Brit. Anem. p. 88,
pl. 1. fig. 3, pl. 2. fig. 5, &c.

Everywhere abundant under stones, and attached to rocks
near low-water mark. In regard to the physiology of the
digestive sac, Mr. Gosse * states that the walls of this chamber
are only separated for the reception of food; but in this species
the mouth often expands, and the digestive cavity dilates so
as to be readily viewed as an open and empty sac. The
ciliary currents course over the lip and into the stomach; so
that minute particles of nutriment might be available, though
by no means necessary.

Fam. 4. Actiniadæ.

Genus 1. ACTINIA, L.

Actinia mesembryanthemum, Ellis & Sol.; Gosse, Brit. Anem.
p. 175, pl. 6. figs. 1–7.

Very common on stones and rocky ledges between tide-
marks.

* Brit. Anem., Introd. p. xvi.

Fam. 9. **Bunodidæ.**

Genus 3. TEALIA, Gosse.

Tealia crassicornis, O. F. Müller; Gosse, Brit. Anem.
p. 209, pl. 4. fig. 1.

The variety *coriacea* (*Actinia coriacea*, Cuvier) is extremely
abundant along the West Rocks at low water, while the other
comes in great profusion and of large size from the deep water
of the bay. A bifid specimen occurred at the Castle rocks.
This species is also found in the stomach of the cod.

Genus 5. STOMPHIA, Gosse.

Stomphia Churchiæ, Gosse, Brit. Anem. p. 222, pl. 8. fig. 5.
Occasionally from deep water.

Fam. 6. **Ilyanthidæ.**

Genus 2. PEACHIA, Gosse.

Peachia hastata, Gosse, Brit. Anem. p. 235, pl. 8. fig. 3.
[Plate II. figs. 5, 6, & 7.]

Thrown ashore on the West Sands after storms in great
numbers, and was thus first found in Britain by Dr. John
Reid, of St. Andrews, who published a description of his
single example in 1848 (Physiological, Anat., and Pathol.
Observations, p. 656, pl. 5. f. 21 & 22): his title (*A. cylin-
drica*) has therefore a prior claim to that of Mr. Gosse. It
occurs also in the stomach of the cod.

Genus 4. EDWARDSIA, De Quatrefages.

Edwardsia callimorpha, Gosse, Brit. Anem. p. 255,
pl. 7. fig. 7.

A variety (Plate II. figs. 1 & 2) was found on the West
Sands after a storm in March, with brown instead of the
usual whitish specks. It is an elongated form inhabiting
sand.

Edwardsia Allmanni, M'I., Proc. Roy. Soc. Ed. 1864–5.
[Plate II. fig. 3.]

From a shallow pool on the West Sands after a storm in
October. It inhabits a distinct case, and can retract its
tentacles and cover them by the external border of the

disk (Plate VII. figs. 1 & 2). The latter is marked by
eight alcyonarian divisions or radii, and has always a
ragged border of the investing sheath. The disk has a pale
brownish colour.

The tentacles (Plate VII. fig. 4) are simple, rather blunt,
pale and translucent, with a white streak in the centre;
the rim of the mouth (fig. 3) is occasionally protruded as a
conical process.

This form exhibited none of the "remarkably vigorous and
spasmodic contractility" ascribed by Mr. Gosse to the family;
for it was comparatively inert.

Edwardsia Goodsiri, M'I., Proc. Roy. Soc. Ed. 1864–5.
[Plate II. fig. 4.]

Found at the same time and place with the former. Tentacles 15, translucent, longer than the diameter of the oral
disk, and not much tapered (Plate VII. fig. 6). A whitish
ring occurs at the tip of each; and from the base a white
spear-head with a transparent centre reaches more than half-
way up (fig. 7). Oral disk streaked with white and brown.
It is somewhat allied to *E. Beautempsii*, De Quatref.*, but is
distinguished by the marks on the tentacles, which in the
latter only have the tip "d'un beau jaune rougeâtre." The
posterior end of the example (fig. 5) was often fixed to the
glass by its ectoderm, which apparently had very minute or
granular suckers.

Swarms of an *Edwardsia* occur in the stomach of the
flounder.

Genus 6. CERIANTHUS, Delle Chiaje.

Cerianthus Lloydii, Gosse, Brit. Anem. p. 268, pl. 7. fig. 8,
and woodcut, p. 269.

Procured at low water from the margin of the East Rocks,
and occasionally thrown on the West Sands after storms. A
splendid specimen from the latter (measuring 7¼ inches long
and as thick as a finger) in February discharged a vast
number of ova after a week's confinement. The majority of
these ovoid bodies were rather coarsely granular; and some
had minute papillæ at one end. No cilia were present; so
that in all probability they were dead. Both examples had

* Ann. des Sc. Nat. 2e sér., Zool. xviii. 1842, p. 60, pl. 1. fig. 1.

the first series of tentacles of the usual brown colour, with
about four faintly marked whitish specks on the inner surface.
The second series were uniformly brown.

Order II. ALCYONARIA.

Fam. Pennatulidæ.

Genus PENNATULA, L.

Pennatula phosphorea, L.; Johnst. Brit. Zooph.
p. 157, fig. 35.

Abundant on muddy ground in deep water, and often
brought up on the lines of the fishermen.

Genus VIRGULARIA, Lamck.

Virgularia mirabilis, L.; Johnst. Brit. Zooph. p. 161, pl. 30.

Occasionally in the stomach of the cod.

Fam. Alcyoniadæ.

Genus ALCYONIUM, L.

Alcyonium digitatum, L.; Johnst. Brit. Zooph.
p. 174, pl. 34.
[Plate VII. figs. 8 to 12.]

Abundant in deep water, as well as in small patches on
rocks and stones between tide-marks. Often thrown in large
quantities on the West Sands after storms, attached to various
submarine structures.

Colony of young *Actinoloba dianthus*.

Subkingdom *MOLLUSCA*.

Section I. MOLLUSCOIDA.

Class I. POLYZOA.

The majority of the Polyzoa come from the deep water of the bay; and, indeed, there are comparatively few to be met with between tide-marks that do not also occur in the former. The minute animals of the calcareous masses so characteristic of many of this group, perform none of those alterations on the surface of the earth which the equally tiny coral-polyps daily effect; yet their workmanship in our northern waters is as regular and beautiful as that fashioned by the latter in the tropical seas. The patterns of the *Lepraliæ*, for instance, excite admiration; and though the apparent resemblance in growth, superficial aspect, and position may suggest to some an analogy between them and the lichens of our rocks and trees, yet it is remote and unable to bear close criticism. It is true it is difficult to assign an exact function to these organisms; but in some cases the calcareous crust of the *Lepraliæ* affords a better hold to many stationary marine animals than the rock itself. Moreover, after heavy-coated forms (like the *Balani*) have reared themselves on this basis, it frequently happens that the original crust is loosened from its attachment, and both fall off together. The coating of *Lepraliæ*, also, may prevent to some extent the disintegration of soft rocks and stones. By removing a portion of bark with an adherent *Balanus* from a submerged thorn-tree, and carefully detaching the former, a fine network of *Lepralia* is found lowest, then the calcareous coating of the *Balanus*; and if the latter has perished, a rough layer of *Cellepora pumicosa* obliterates all trace of it from without.

The Cheilostomatous Polyzoa are fairly represented; and several, e. g. *Flustra* and *Gemellaria*, occur in vast quantities attached to stones, shells, and corallines on the West Sands after storms. The majority are common to the eastern shores, the west, and the extreme north and south, as shown in the valuable catalogues of Messrs. Alder, Couch, Hincks, and Norman. Many species will doubtless yet be found at St. Andrews—though at present they appear to be confined to the other areas, which have been more thoroughly investigated by observers specially skilled in this department. *Bicellaria ciliata* and *Bugula purpurotincta* seem to be more common at St. Andrews than in Shetland, the latter form being especially abundant and fine, and apparently taking the place of the *B. plumosa* of the southern shores; *Menipea ternata* and *Bugula Murrayana* are likewise in considerable profusion and in fine condition; while the southern *Flustra chartacea* is wholly absent. The species of the Membraniporidæ, perhaps, are more abundant in Shetland; and the *Lepraliæ* are decidedly more numerous there and in the extreme south. Amongst the more conspicuous forms we notice the entire absence of *Lepralia Pallasiana*, so common in the extreme west and south, and of the characteristic *L. innominata* and *L. figularis* of the latter. The Celleporidæ are abundant, but the species are few. *Cellepora avicularis* is exceptionally rich, according to Mr. Hincks; and the same high authority in this department states that the sole specimen of *Eschara Skenei* is fine.

The Cyclostomatous forms are not numerous; but all the examples are abundant; and the same may be said of the Ctenostomata. The late Dr. John Reid mentions *Vesicularia spinosa* as growing near low-water mark; but I have not been successful in finding it. The Zetlandic *Hornera* and the rich tufts of *Amathia lendigera*, so plentiful in the south, are altogether absent.

On the whole it would appear that the Hebridean, Zetlandic, and southern waters furnish a richer field for the Polyzoa than our eastern shores, not only as regards the number of species, but the condition and size of the specimens. I need only allude, for instance, to the luxuriance of the branching Celleporidæ and Reteporæ of the Hebrides and Shetland, and the extraordinary beauty and profusion of the Escharidæ and Lepraliæ, and indeed of the whole group, in the extreme south and in the Channel Islands, both between tide-marks and on the shell-banks around.

G

The arrangement followed is that of Mr. Busk in his
accurate, well-known, and beautifully-illustrated 'Catalogue;'
and I have further derived great assistance from the valuable Catalogue of the Zoophytes of Northumberland and
Durham by the lamented Joshua Alder, and the extensive
Zetlandic lists by the Rev. A. M. Norman. I have also to
thank Mr. Hincks for his kindness in revising the list and
making several additions, and to acknowledge the information
derived from his careful and original Catalogue of the southern
forms.

Order GYMNOLÆMATA.

Suborder CHEILOSTOMATA.

Family Salicornariadæ.

Genus CELLARIA, Lamarck.

Cellaria farciminoides, Ellis & Solander ; Busk, Catal. p. 16,
    pl. 64. f. 1–3, pl. 65 (bis). f. 5.

Attached to the roots of *Flustra foliacea* and other corallines ;
abundant in deep water.

Family Cellulariadæ.

Genus MENIPEA, Lamx.

Menipea ternata, Ellis & Solander ; Busk, Catal. p. 21,
    pl. 20. f. 3–5.

Fine tufts on *Sertularia filicula* and other corallines from
the deep water of the bay.

Genus SCRUPOCELLARIA, Van Beneden.

Scrupocellaria scruposa, L. ; Busk, Catal. p. 25, pl. 22.
    f. 3 & 4.

Abundant under stones between tide-marks, and ranging to
deep water. In October many are marked with the reddish
orange ova ; there are also numerous brownish black specks
on these specimens.

Genus CANDA, Lamx.

Canda reptans, Pallas ; Busk, Catal. p. 26, pl. 21. f. 3 & 4.

Found by Dr. John Reid near low-water mark (Anat. and
Pathol. Observat. p. 602).

Family **Scrupariadæ**.

Genus SCRUPARIA, Oken.

Scruparia chelata, L.; Busk, Catal. p. 29, pl. 17. f. 2.

Common on *Ceramium rubrum*, *Sertularia pumila*, and other algæ and zoophytes between tide-marks.

Genus HIPPOTHOA, Lamx.

Hippothoa catenularia, Jameson; Busk, Catal. p. 29, pl. 18. f. 1 & 2.

On stones and shells from deep water; less common than the following species.

Hippothoa divaricata, Lamx.; Busk, Catal. p. 30, pl. 18. f. 3 & 4.

On stones and shells from deep water; abundant.

Family **Gemellariadæ**.

Genus GEMELLARIA, Sav.

Gemellaria loricata, L.; Busk, Catal. p. 34, pl. 45. f. 5 & 6.

Abundant in deep water, and thrown on shore in masses after storms.

Family **Bicellariadæ**.

Genus BICELLARIA, De Blainville.

Bicellaria ciliata, L.; Busk, Catal. p. 41, pl. 34.

Frequent on stones and shells from the coralline ground.

Genus BUGULA, Oken.

Bugula flabellata (J. V. Thompson, MS.), Gray; Busk, Catal. p. 44, pls. 51 & 52.

On *Flustra foliacea* from deep water; rather rare.

Bugula avicularia, Pallas; Busk, Catal. p. 45, pl. 53.

From the coralline ground, on *Flustra truncata*; not common.

G 2

Bugula purpurotincta, Norman, Quart. Journ. Micr. Sci.
n. s. vol. viii. p. 219.

Abundant on the same ground, attached to shells.

Bugula Murrayana, Bean; Busk, Catal. p. 46, pl. 59.

Plentiful on the beach after storms, and at all times from
the coralline ground.

Family Flustridæ.

Genus FLUSTRA, L.

Flustra foliacea, L.; Busk, Catal. p. 47, pl. 55. f. 4 & 5,
pl. 56. f. 5.

Very abundant on the sands after storms.

Flustra truncata, L.; Busk, Catal. p. 48, pl. 56. f. 1 & 2,
pl. 58. f. 1 & 2.

Common in the laminarian and coralline zones.

Genus CARBASEA, Gray.

Carbasea papyrea, Pallas; Busk, Catal. p. 50, pl. 50. f. 1–3.

After storms, and from the fishing-boats; not abundant.

Family Membraniporidæ.

Genus MEMBRANIPORA, De Blainville.

Membranipora membranacea, L.; Busk, Catal. p. 56, pl. 68. f. 2.

Abundant on the fronds of *Laminaria digitata* and other
algæ.

Membranipora pilosa, L.; Busk, Catal. p. 56, pl. 71.

Very common on the stems and fronds of *Delesseriæ, Laminariæ*, and other seaweeds between and beyond tide-marks,
and on shells and stones from the coralline ground.

Membranipora Flemingii, Busk; Catal. p. 58, pl. 61. f. 2,
pl. 84. f. 4–6, pl. 104. f. 2–4.

Common on stones and shells from the coralline ground.

Membranipora Lacroixii, Sav.; Busk, Catal. p. 60, pl. 69,
pl. 104. f. 1.

On the inner surface of a valve of *Cyprina islandica* from
deep water.

Membranipora spinifera, Johnst.; Alder, Catal. Zooph. p. 53,
pl. 8. f. 2, 2 a.

Abundant on the under surface of stones between tide-
marks.

Membranipora Dumerillii, Aud.; Alder, Catal. Zooph. p. 56,
pl. 8. f. 5.

Not uncommon on bivalves from deep water. As Mr. Alder
observes, it may occasionally be seen in company with *M.
Flemingii*.

Membranipora unicornis, Flem.; Alder, Catal. Zooph. p. 56,
pl. 8. f. 6.

On bivalves from deep water; not very common.

Membranipora craticula, Alder; Catal. Zooph. p. 54, pl. 8. f. 3.

Occasionally in deep water.

Genus FLUSTRELLA, Gray.

Flustrella hispida, Fab.; Johnst. Brit. Zooph. p. 363,
pl. 66. f. 5.

Abundant on the stems of Fuci and other seaweeds, and on
stones between tide-marks.

Genus LEPRALIA, Johnst.

Lepralia Brongniartii, Aud.; Busk, Catal. p. 65,
pl. 81. f. 1-5.

Rather plentiful on laminarian roots thrown on shore after
storms. Often forms a basis for other growths, and may be
seen on their under surface when detached from seaweed or
rock.

Lepralia reticulata, J. Macgillivray ; Busk, Catal. p. 66,
    pl. 90. f. 1, pl. 93. f. 1 & 2, pl. 102. f. 1.

Not uncommon in the siphons and inside the mouth of
Fusus antiquus, and also on *Cardium echinatum* from deep
water.

Lepralia concinna, Busk, Catal. p. 67, pl. 99.

Very abundant on stones and shells from the coralline
ground.   A well-marked variety, with perforations round the
cells, is not uncommon.

Lepralia verrucosa, Esper; Busk, Catal. p. 68, pl. 87. f. 3 & 4,
    pl. 94. f. 6.

Occurs rather abundantly on the roots of *Laminaria digitata*
and on stones near low-water mark.

Lepralia unicornis, Johnst. ; Brit. Zooph. p. 320, pl. 57. f. 1.

A common littoral species, everywhere abundant, and in
large patches on the under surface of stones.   The colours
vary, probably in some cases from the ova.

Lepralia spinifera, Johnst. ; Busk, Catal. p. 69, pl. 76. f. 2 & 3.

Very common on the under surface of stones near low-
water mark.

Lepralia trispinosa, Johnst. ; Busk, Catal. p. 70,
    pl. 85. f. 1 & 2, pl. 98, pl. 102. f. 2.

Abundant on stones and shells from the coralline zone.

Lepralia coccinea, Abildgaard ; Busk, Catal. p. 70, pl. 88.

On sandstone, shale, and laminarian roots from the East
Rocks, and on shells from deep water.   Also found by Prof.
J. Reid.   Rare as contrasted with its profusion on our south-
ern shores.

Lepralia linearis, Hassall ; Busk, Catal. p. 71, pl. 89. f. 1–3.

Common on shells and stones from deep water.

Lepralia ciliata, Pallas ; Busk, Catal. p. 73, pl. 74. f. 1 & 2, pl. 77. f. 3, 4, 5.

Occasionally on the under surface of stones near low-water mark ; more frequently on stones and shells from the coralline ground.

Lepralia variolosa, Johnst. ; Busk, Catal. p. 75, pl. 74. f. 3, 4, 5, pl. 75.

On shells and stones from deep water ; not uncommon.

Lepralia nitida, Fab. ; Busk, Catal. p. 76, pl. 76. f. 1.

Abundant both between tide-marks and in deep water, on stones and shells. The spines are in general less developed than in those from the Channel Islands.

Lepralia annulata, Fab. ; Busk, Catal. p. 76, pl. 77. f. 1.

Instead of being partial to the laminarian blades, as on the west coast, this species is not uncommon on the under surface of stones between tide-marks, generally in small patches ; and also occurs on shells and stones from deep water. Some dried specimens are of a pinkish colour.

Lepralia Peachii, Johnst. ; Busk, Catal. p. 77, pl. 82. f. 4, pl. 97.

Common on stones near low-water mark, and on stones and shells from deep water.

Lepralia ventricosa, Hassall ; Busk, Catal. p. 78, pl. 82. f. 5 & 6, pl. 83. f. 5, pl. 91. f. 5 & 6.

Not uncommon on stones and shells from deep water.

Lepralia punctata, Hassall ; Busk, Catal. pl. 90. f. 5 & 6, pl. 92. f. 4, pl. 96. f. 3.

Everywhere abundant on the under surface of stones in pools and elsewhere near low-water mark, and also on shells and stones from deep water.

Lepralia Malusii, Aud. ; Busk, Catal. p. 83, pl. 103. f. 1–4.

Not uncommon on shells and stones from deep water.

Lepralia granifera, Johnst.; Busk, Catal. p. 83, pl. 77. f. 2,
pl. 95. f. 6 & 7.

Abundant on the under surface of stones near low-water mark in considerable patches. The aspects of the old and new cells differ much. The new cells glisten like those of *L. hyalina*, have a number of opaque white granules, a D-shaped aperture, and a distinct mucro; the transverse wrinkles of the cells are also apparent; and in some very new ones the granules are also glistening and hyaline, and show the perforations. In the old cells the walls are opaque, whitish, or yellowish, the granules still more opaque, perhaps larger, but less defined and beautiful.

Lepralia hyalina, L.; Busk, Catal. p. 84, pl. 82. f. 1–3,
pl. 95. f. 3–5, pl. 101. f. 1 & 2.

Common on laminarian roots and stems, on *Delesseria* and other algæ, and on stones near and beyond low-water mark.

Family Celleporidæ.

Genus CELLEPORA, Fab.

Section A. *Incrusting, adnate.*

Cellepora pumicosa, L.; Busk, Catal. p. 86, pl. 110. f. 4–6.

Very abundant on stones, shells, zoophytes, and seaweeds—generally from deep water.

Cellepora avicularis, Hincks; Catal. Zooph. Devon,
Ann. & Mag. Nat. Hist. 3rd ser. ix. p. 304.

Occasionally on zoophytes.

Section B. *Erect, branching.*

Cellepora ramulosa, L.; Busk, Catal. p. 87, pl. 109. f. 1–3.

Attached to the stems of zoophytes &c. in deep water; common.

Cellepora dichotoma, Hincks, Catal. Zooph., *loc. cit.* p. 305.

On zoophytes; abundant and fine.

Family Escharidæ.

Genus ESCHARA, Ray.

Eschara Skenei, Ellis & Sol. ; Busk, Catal. p. 88, pl. 122.

A remarkably beautiful specimen on *Cyprina islandica* from the coralline ground.

Suborder CYCLOSTOMATA, Busk.

Family Tubuliporidæ, Johnst.

Genus TUBULIPORA, Lamarck.

Tubulipora serpens, L.; Johnst. Brit. Zooph. p. 275, pl. 47. f. 4-6.

On zoophytes and shells from deep water ; very abundant and characteristic.

Genus ALECTO, Lamx.

Alecto granulata, M.-Ed.; Johnst. Brit. Zooph. p. 280, pl. 49. f. 1 & 2.

On stones and shells from deep water; not rare.

Family Diastoporidæ, Busk.

Genus DIASTOPORA, Lamx.

Diastopora obelia, Flem.; Johnst. Brit. Zooph. p. 277, pl. 47. f. 7 & 8.

On shells and stones from deep water.

Genus PATINELLA, Gray.

Patinella patina, Lamarck ; Hincks, Catal. Zooph., *loc. cit.* p. 468.

Abundant on corallines and shells from deep water, especially on *Mytilus modiolus.*

H

Genus HETEROPORELLA, Busk.

Heteroporella hispida, Flem.; Hincks, *loc. cit.* p. 469.
On stones and shells from deep water; rather rare.

Family **Crisiadæ**.

Genus CRISIA, Lamx.

Crisia eburnea, L.; Johnst. Brit. Zooph. p. 283,
pl. 50. f. 3 & 4.

On the under surface of stones between tide-marks, often
with many parasitic hydroids and Confervæ, and on zoophytes
and seaweeds from deep water.   Abundant.

Suborder **CTENOSTOMATA**, Busk.

Family **Alcyonidiadæ**.

Genus ALCYONIDIUM, Lamx.

Alcyonidium gelatinosum, Pallas; Johnst. Brit. Zooph.
p. 358, pl. 68. f. 1–3.

On stones and bivalve shells from deep water; common.

Alcyonidium hirsutum, Flem.; Johnst. Brit. Zooph. p. 360,
pl. 69. f. 1 & 2.

Abundant on seaweeds near and beyond low-water mark.

Alcyonidium parasiticum, Flem.; Johnst. Brit. Zooph.
p. 362, pl. 68. f. 4 & 5.

Frequent on the stems of zoophytes from deep water; very
characteristic.

Genus ARACHNIDIA, Hincks.

Arachnidia hippothooides, Hincks, Ann. & Mag. Nat. Hist.
3rd ser. ix. p. 471, pl. 16. f. 2.

On *Ascidia sordida* from deep water; in abundance.

Family **Vesiculariadæ**.

Genus Vesicularia, J. V. Thompson.

Vesicularia spinosa, L.; Johnst. Brit. Zooph. p. 370,
pl. 72. f. 1–4.

Found near low-water mark by the late Prof. John
Reid.

Genus Bowerbankia, Farre.

Bowerbankia imbricata, Adams; Johnst. Brit. Zooph.
p. 377, pl. 72. f. 5 & 6.

Abounds on the under surface of stones, on the stems and
branches of littoral zoophytes, and on the tests of *Cynthia
grossularia* under shelving rocks. Also found by Prof. J.
Reid.

Order Phylactolæmata.

Suborder **PEDICELLINEA**.

Family **Pedicellinidæ**.

Genus Pedicellina, Sars.

Pedicellina echinata, Sars; Johnst. Brit. Zooph. p. 382,
pl. 70. f. 5.

On the branches of *Ceramium rubrum* and other littoral
algæ and zoophytes; abundant.

Class II. TUNICATA.

Comparatively few Ascidians have been procured; indeed
the department is in such a condition at present (as to specific
identification) that a much greater amount of time would have
been required for their elucidation than was available. The
late Mr. Joshua Alder most kindly looked over the collection,
and named those requiring identification in his usual con-
scientious manner; and it is to be hoped that the work on

these forms by him and the late Mr. Hancock (one of the
best minute anatomists this country has produced) will soon
be published. The most abundant simple form is the *Ascidia
sordida* of Alder and Hancock, which is thrown by storms
on the West Sands in large numbers, attached to seaweeds, sticks, shells, and other objects. *A. intestinalis* is also
procured in this manner as well as between tide-marks;
Pelonaia corrugata and *Molgula arenosa*, A. & H., affect deep
water only, and rarely occur during storms. The compound
forms are common under stones between tide-marks and in
the laminarian region : but much yet remains to be done in
this respect at St. Andrews. Though Ascidians on the exposed parts of the east coast of Scotland are for the most part
rare in the laminarian region and between tide-marks, they
are common in still muddy waters on the west coast and in the
Hebrides, and in water which cannot but be slightly brackish,
as at the head of Loch Portan near Lochmaddy, where they
are both abundant and large; they are also numerous and
large between tide-marks at Herm and in the rich waters
around the Channel Islands, as well as in the Zetlandic
voes.

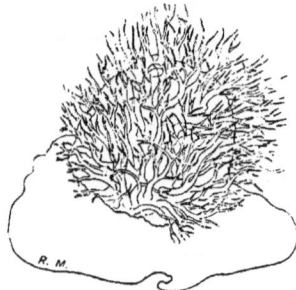

Cynthia echinata, L.   (From the Minch.)

Family **Botryllidæ**.

Genus LEPTOCLINUM, M. Edwards.

Leptoclinum durum, M. Ed.; Forbes & Hanley, Brit. Mollusca,
    i. p. 17 (as *L. aureum*, a misprint).

Common under stones in rock-pools between tide-marks.
Dull yellowish white, with white specks from stellate calcareous crystals.

Leptoclinum punctatum, Forbes; F. & H. Brit. Moll. i.
            p. 18.

Not uncommon under stones between tide-marks.

Genus BOTRYLLUS, Gærtner.

Botryllus Schlosseri, Pallas; F. & H. Brit. Moll. i.
        p. 19, pl. A. fig. 7, pl. B. fig. 7.

Occasionally under stones between tide-marks. The red
spot in the centre is not very visible in these specimens. On
tearing, a dark brownish digestive system appears.

Botryllus polycyclus, Sav.; F. & H. Brit. Moll. i.
            p. 21.

Frequent near low-water mark on the under surface of
stones, on Fuci and *Corallina officinalis*.

Genus BOTRYLLOIDES, M. Edwards.

Botrylloides Leachii, Sav.; F. & H. Brit. Moll. i. p. 23.

Common in the laminarian region attached to seaweeds.

Numerous other species of *Botrylloides* and a *Didemnum* are
common under stones in the rock-pools.

Genus PARASCIDIA, Alder.

Parascidia Flemingii, Alder, Ann. & Mag. Nat. Hist.
        1863, xi. p. 172.
        [Plate IX. fig. 3.]

Occasionally on laminarian roots near low water. Mr.

Alder was of opinion that the drawing represented a young form of this species. It consisted of cylindrical animals with a transparent investment. On the summit of each are several long-ovate reddish orange structures marked with yellowish white grains, showing at the free extremity an oral aperture surrounded by eight small papillæ.

Family Clavelinidæ.

Genus CLAVELINA, Sav.

Clavelina lepadiformis, O. F. Müller; F. & H. Brit. Moll. i. p. 26.

[Plate II. fig. 9.]

Occasionally between tide-marks.

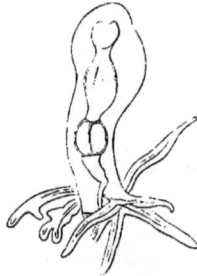

Clavelina lepadiformis, O. F. Müller.

Family Ascidiadæ.

Genus ASCIDIA, Baster.

Ascidia intestinalis, L.; F. & H. Brit. Moll. i. p. 31.

Abundant on the roots of tangles thrown on the West Sands

after storms, and also under stones between tide-marks. Evinces considerable and spasmodic muscular contractions.

Ascidia sordida, Ald. & Hanc.

[Plate IX. figs. 1 & 2.]

Very plentiful in the deeper water attached to stones, shells, sticks, and seaweeds. One has placed itself on the anterior end of an empty tube of *Pectinaria belgica* and quite filled it up (Plate IX. figs. 1 & 2). An elongated (club-shaped) variety is not uncommon.

Ascidia depressa, Ald. & Hanc.

Not uncommon on the under surfaces of large stones in tide-pools. In November specimens are often loaded with a pinkish-white creamy fluid, which appears to be made up chiefly of ova. The cellular border of each ovum is faintly greenish by transmitted light.

Genus MOLGULA, E. Forbes.

Molgula arenosa, Ald. & Hanc.; Alder, *op. cit.* p. 160.

Abundant in deep water, and in the stomach of the cod and haddock.

Genus CYNTHIA, Sav.

Cynthia, n. sp.

A nodulated Ascidian like a raspberry or small bramble occurred on the West Sands after a storm. Mr. Alder stated that it was not *C. morus*, but a species unknown to him.

Cynthia grossularia, Van Beneden; F. & H. Brit. Moll. i. p. 40.

Very common on the roofs of rocky ledges, between tide-marks, where it becomes incrusted by many parasites. The development of this species is easily observed.

Genus PELONAIA, Forbes & Goodsir.

Pelonaia corrugata, Forbes & Goodsir ; F. & H. Brit. Moll.
i. p. 43, pl. E. f. 4.

[Plate II. fig. 10.]

Frequent in deep water, and occasionally in the stomachs
of the cod and haddock.

Section II. MOLLUSCA (PROPER).

The Mollusca are chiefly procured by dredging, examination between tide-marks, or the deep-sea lines of the fishermen, though certain storms sometimes strew the sands with many species in great profusion. Not a few of the rarer forms are found in the stomachs of fishes, such as the cod, haddock, and flounder. The remarks on the class may be arranged in three divisions, founded on the economical value, peculiar habits, and rarity.

By far the most important species in the first group is the common mussel (*Mytilus edulis*), which occurs in vast numbers in the form of mussel-" beds " on muddy flats, chiefly situated on the right bank of the estuary of the river Eden. Attached to stones, sticks, and to each other, these shell-fish luxuriate in abundance of Diatomaceæ, Infusoria, and other minute forms of animal and vegetable life. From their special value as bait the city derives a considerable annual revenue; and if the wise protection only lately enforced were supplemented by skilful mussel-farming, great increase in revenue might be anticipated. Shell-fish, like other animals in civilized parts, cannot survive constant inroads without special restrictions. Multitudes of the young mollusks, moreover, are found incrusting the poles for the salmon-nets on the West Sands, and the rocks, stones, and tangle-roots elsewhere; but they do not attain a large size, apparently from overcrowding and the want of congenial food, which the purer water seems incapable of supplying. This species takes the place of the horse-mussel ("yoags") of the Zetlandic fishermen, and the worms of those in the Channel Islands. It is seldom eaten by the natives. The edible cockles, again, abound on the sandy flats near the entrance of the Eden into the sea, and are occasionally sold as food, though of late years their scarcity has rendered their appearance less frequent in the market. Periwinkles and limpets are constantly gathered for similar purposes. The

1

only uses to which some of the other mollusks are applied are
in the amateur manufacture of ornaments, such as shell pin-
cushions from various univalves and bivalves, bracelets from
Nassa incrassata and *Trochus cinerarius*, after the latter has
had its nacreous layer exposed by an acid.

The rock-boring shell-fish are five in number, though only
one exercises any great influence on the disintegration of the
rocks—viz. *Pholas crispata*, which often swarms in the shale
and sandstone, and takes the place of the *Pholas dactylus* of
the chalk rocks of the south. *Pholas candida* occurs too rarely
to require special mention in this respect; and the same may
be said of *Saxicava rugosa*. The excavations of *Patella vulgata* and *Chiton fascicularis* in sandstone show that no special
boring-organ is necessary for this purpose. The latter species
sometimes scoops out considerable cavities in sandstone, in
which it reposes. The only wood-borer is *Xylophaga dorsalis*.

In taking, under this head, a general survey of the boring
forms, it is found that they belong to at least three invertebrate
subkingdoms, viz. the Protozoa, Mollusca, and Annulosa. In
the first are boring sponges; in the second, Bryozoa and various
mollusks; in the third, sea-urchins, gephyreans, annelids, and
cirripedes.

The sponges appear to bore only into calcareous substances,
such as shells and limestone. The Bryozoa perforate shells;
the mollusks proper, limestone, sandstone, aluminous shale,
gneiss and other rocks, wood, wax, shells, &c.; the annelids
tunnel shells and rocks of various kinds; the sea-urchins calcareous rocks, gneiss, granite, and other rocks; the gephyreans
and cirripedes shells and limestone. Good opportunities are
afforded at St. Andrews for studying the boring-action of the
mollusks and annelids. *Pholas crispata* and *Leucodore ciliata*
are equally abundant, and must exercise as much influence
on the disintegration of the sandstone and shale between tide-
marks as the boring sponges amongst the shells in deep water;
while *Patella*, *Chiton*, *Saxicava*, and *Dodecaceria* are also
frequent.

The theories which have been promulgated to explain the
mode by which these various creatures perforate materials so
diverse may be ranged round two great centres, viz. the

chemical and the *mechanical**. The advocates of the former seem to take for granted that the borings occur chiefly in calcareous substances; and with propriety, therefore, they make their solvent an acid. It is clear, however, that this agency is unable to account for the abundant perforations in media totally impervious to such action. Moreover no trace of acid is found in many borers; and though present in some, as in *Sabella saxicava* and *Pholas*, it is likewise characteristic of other marine animals which do not bore; and it is purely hypothetical at present to bring in the aid of the carbonic dioxide derived from sea-water, for the same reason.

The mechanical theory, again, supposes that the animals perforate by means of shells or gritty particles in the case of mollusks, teeth in the sea-urchins, bristles in the annelids, horny processes in certain cirripedes and gephyreans; but we are left in doubt concerning the extensive and wonderful perforations of the sponges, those of the Bryozoa, and the rest of the cirripedes. If, however, we regard the "macerating" theory as a modification of this, certain of the difficulties will be overcome. The grains of wood, however, found in the stomachs of *Teredo*, are interesting in this respect.

The whole subject of the boring of marine animals, indeed, is much in want of further elucidation; and it is hard to believe that the same *modus operandi* is employed by all. In conclusion, the theories may be arranged under the following heads (for all the subkingdoms), after Forbes and Hanley and Gwyn Jeffreys :—

1. That in the shell-fish the perforations are made by rotations of the valves, like augers (Bonanni, Adanson, Born, J. E. Gray, Dr. Fleming, Osler, Forbes and Hanley, Cailliaud, Robertson, &c.); in the sea-urchins, by the teeth (Cailliaud).

   This theory is not supported by an examination of the perforations of the sponges, Bryozoa, those of the annelids, gephyreans, and cirripedes, nor by a comparison of the

* We do not here allude to the boring by jaws or tongue (*e. g.* in *Limnoria* and *Trochus*).

I 2

shells and tunnels of the mollusks themselves. The epidermis of the latter in each case would likewise suffer.

II. That the holes are made by rasping effected by siliceous particles on the foot of the mollusk (Hancock), by grains of silex from the exterior (Bryson), by the foot in some way (Dr. Fischer), by chitine in the cirripedes (Darwin) and gephyreans and the bristles of the annelids.

This explanation is not borne out by the case of the sponges, by that of the Bryozoa, and certain cirripedes ; moreover such siliceous particles are rare in boring mollusks.

III. That the excavations are due to ciliary currents, aided by rasping (Garner).

The currents may assist, but seem to be insufficient to account for the borings of any group.

IV. That the perforations are produced by a chemical solvent : Gray, Osler (for *Saxicava*), Drummond, Cailliaud, Mantell, Thoreau, Reeve, Bouchard-Chantereaux, Spence Bate, Darwin (for *Verruca*), E. R. Lankester, and Parfitt.

This will not explain the borings in wood, aluminous shale, gneiss, granite, sandstone, and wax. It is interesting, however, as my friend Mr. Ray Lankester has specially shown, that shells and calcareous rocks are much affected by burrowing marine animals.

V. That the borings are the result of a secreted solvent and rasping action (Thompson and Necker).

It seems improbable that the solvent should vary with the nature of the rocks attacked.

VI. That the perforations are caused by a macerating or simple solvent action of the foot in mollusks (Sellius, Deshayes, and Gwyn Jeffreys).

It is doubtful if this would be supported by the action of the sponges, Bryozoa, annelids, gephyreans, and cirripedes. The striæ in certain of the tunnels of the shell-fish are also somewhat at variance with this notion.

The most interesting species in regard to rarity are *Lima subauriculata* and *L. Loscombii*, which come from deep water, the characteristic *Lima hians* of our southern and western shores being absent. A worn valve of *Isocardia cor* found on the West Sands is purely accidental. *Tellina pusilla* and *Psammobia tellinella* are uncommon at St. Andrews. Amongst univalves, *Trichotropis borealis*, *Pleurotoma Trevelyana*, *Aplysia punctata*, and *Philine pruinosa* are noteworthy. The smaller univalves, such as *Rissoæ* and their allies, are much less numerous in species than on the southern and western shores, the absence of *Trochus umbilicatus* being especially diagnostic when contrasted with the latter. The Nudibranchs are well represented at all seasons; and the individuals in the majority of the species are numerous. *Ommatostrephes* and *Loligo* amongst the cuttles often occur in great profusion on the West Sands after storms.

On the whole the species are northern, and stand in strong contrast to the molluscan fauna of the western shores, where *Thracia convexa*, *Tapes decussatus*, *Pecten varius*, var. *nivea*, *Teredo megotara* and *T. norvegica*, *Fissurella*, *Trochus umbilicatus* and *T. zizyphinus* in the littoral zone, and the abundance of *T. magus* and *T. tumidus* in the laminarian, *Phasianella*, *Akera bullata*, *Elysia*, swarms of large and small *Rissoæ*, and the pelagic *Ianthina* form conspicuous features of the marine fauna, just as the hosts of *Bulimus acutus* do on the sandy fields of Killipheder and other parts of the extreme west. Still more evident is the contrast with the rich southern species that cluster round the Channel Islands—such as the finely developed pectens, oysters, and *Anomiæ*, and the appearance of the former between tide-marks (*P. varius*), besides *Mytilus barbatus* (which takes the place of the bearded varieties of our *Mytilus modiolus*) in obscure crevices in the littoral zone, the frequency of *Arca tetragona* in fissures of the rocks, *Galeomma* on the under surface of stones in tide-pools at Herm, the boring *Gastrochæna* in shells, and the abundance of *Haliotis*, *Pandora*, *Venus verrucosa* and *V. ovata*, *Mactra glauca*, the *Psammobiæ*, and the "angel's wings" (*Lima*), which when disturbed flit with their brilliant orange fringes so nimbly in the tide-pools. Amongst univalves, again, the large size and abundance of *Chiton discrepans*, *Fissurella*, *Emarginula*, *Murex erinaceus*, *Aplysia punctata*, *Eulima polita*, *Trochus lineatus*, *Cerithium* and *Cerithiopsis*, and the predatory and cunning cuttles

(*Octopus*) between tide-marks are noteworthy; while in the surrounding water are the rare prizes *Triton nodifer*, *T. cutaceus*, *Cardium papillosum*, *Argiope decollata*, and other forms which, with the foregoing, are thrown in such profusion on the shell-beach at Herm—e. g. *Calyptræa chinensis*, *Trochus Montacuti*, and *Murex aciculatus*. The fine *Pinna rudis*, though it has been found by Dr. Howden off Montrose, is also entirely absent at St. Andrews. Neither do we find the swarms of *Trochus helicinus* and *T. grœnlandicus*, *Trichotropis borealis*, and their allies amongst the tangle-roots, as in Shetland, nor the *Terebratula*, *Lyonsia*, *Lepeta*, *Puncturella*, *Trochus amabilis*, the *Jeffreysiæ*, *Columbella haliæeti*, *Pleurotoma nivalis*, *P. carinata*, *Scaphander librarius*, *Philine angulata* and *P. nitida*, *Rossia papillifera*, the *Clios*, and the half hundred new British forms discovered by Dr. Gwyn Jeffreys in his frequent cruises in the surrounding waters. The great beds, also, of *Mytilus modiolus* (called "yoags"), which occur in from 3 to 15 fathoms near the shore in many parts of the Zetlandic seas, present an interesting contrast. It is this mussel (esteemed but a coarse bait at St. Andrews) which is extensively employed by the Shetlanders; and in its collection many rare invertebrates are found amongst the roots of the tangles and stones, which with the mussels form huge muddy masses. The old ten-toothed "dreg" noticed by the accomplished Prof. Edward Forbes is still the chief instrument in procuring the shell-fish, and is much more serviceable to the zoologist on such ground than the dredge. In the figure given by the facile pencil of the great naturalist the rope is attached to the eye of the dreg; but in modern times the fishermen more correctly attach it to the head of the instrument (after the manner of the ordinary dredge), and fix the rope at the eye of the dreg by a piece of spun yarn; so that if the dreg gets entangled the

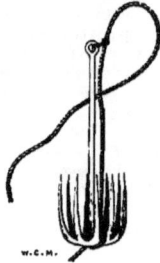

spun yarn gives way, and the rope pulls the head of the dreg backwards, and disengages the teeth from the tangles and stones. In transverse section the teeth form a truncated ellipse round the central iron rod.

The nomenclature adopted is that of Dr. Gwyn Jeffreys in his valuable work on the British Mollusca; and I am specially indebted to him for his great courtesy in frequently aiding me in doubtful cases, and also carefully investigating shell-débris containing minute species, which otherwise might have been overlooked.

Class CONCHIFERA.

Order LAMELLIBRANCHIATA.

Fam. 1. **Anomiidæ**, Gray.

Genus ANOMIA, L.

Anomia ephippium, L. Jeffreys, Brit. Moll. ii. p. 31, v. pl. 20. f. 1, 1 *a*–1 *e*.

Not uncommon in the débris of the fishing-boats.

Anomia patelliformis, L. *Op. cit.* ii. p. 34, v. pl. 20. f. 2, 2 *a*–2 *c*.

Common in the same region, as well as between tide-marks.

Fam. 2. **Ostreidæ**, Broderip.

Genus OSTREA, L.

Ostrea edulis, L. *Op. cit.* ii. p. 38, v. pl. 21.

Living examples are rare. The "rock" variety with purplish streaks, however, is occasionally found at the East Rocks on the under surface of stones in pools near low water.

Fam. 3. **Pectinidæ**, Lamarck.

Genus 1. PECTEN, Pliny.

Pecten pusio, L. *Op. cit.* ii. p. 51, v. pl. 22. f. 1 & 1 *a*.

Common; the living specimens come from the deep water

of the bay, chiefly attached to bivalves. Worn valves are abundant in the gravel at the East Rocks.

Pecten opercularis, L.   *Op. cit.* ii. p. 59, v. pl. 22. f. 3 & 3 a.

Frequently brought by the fishing-boats, and thrown on the beach after storms.

Pecten tigrinus, O. F. Müller.   *Op. cit.* ii. p. 65, v. pl. 23.
f. 2 & 2 a.

Perfect specimens from the coralline ground and the stomachs of haddocks and flounders; single valves on the beach in gravel and after storms.

Pecten similis, Laskey.   *Op. cit.* ii. p. 71, v. pl. 23. f. 5.

Frequent in the stomachs of flounders and haddocks; more rarely procured from the coralline ground.

Pecten maximus, L.   *Op. cit.* ii. p. 73, v. pl. 24.

Occasionally brought up on the deep-sea lines of the fishermen.

Genus 2. LIMA, Bruguière.

Lima subauriculata, Mont.   *Op. cit.* ii. p. 82, v. pl. 25. f. 3.

Not common; from the deep water of the bay.

Lima Loscombii, G. B. Sowerby.   *Op. cit.* ii. p. 85,
v. pl. 25. f. 4.

Single valves occasionally appear in the fishing-boats; perfect specimens are found in the stomach of the cod.

Fam. 5. Mytilidæ, Fleming.

Genus 1. MYTILUS, L.

Mytilus edulis, L.   *Op. cit.* ii. p. 104, v. pl. 27. f. 1.

Constituting by their vast numbers a most important mussel-bed at the estuary of the river Eden. Multitudes of the young animals, besides, form a coating to the posts of the salmon-nets, to rocks, stones, and tangle-roots in various places.

Mytilus modiolus, L. *Op. cit.* ii. p. 111, v. pl. 27. f. 2.

Frequently thrown ashore after storms, and brought by the fishermen from deep water. Monstrosities and varieties are common; and there is no shell so prolific in parasitic or commensalistic growths; pea-crabs and pearls are common in the interior. Young forms (bearded) occur in chinks of the rocks between tide-marks.

Genus 2. MODIOLARIA, Beck.

Modiolaria marmorata, Forbes. *Op. cit.* ii. p. 122, v. pl. 28. f. 1.

Abundant in the tests of *Ascidia sordida*, and sometimes found in a free condition on the West Sands after storms.

Modiolaria discors, L. *Op. cit.* ii. p. 126, v. pl. 28. f. 3.

Occasionally attached to the roots of Fuci near low water, and to the top-shaped fronds of *Himanthalia lorea*.

Modiolaria nigra, Gray. *Op. cit.* ii. p. 128, v. pl. 28. f. 4.

Fine specimens occur in the deep water of the bay, and also in the stomachs of cod.

Genus 3. CRENELLA, Brown.

Crenella decussata, Montagu. *Op. cit.* ii. p. 133, v. pl. 28. f. 6.

Not rare in the stomachs of cod and haddocks, though perhaps swallowed in the first instance by other fishes.

Fam. 6. Arcidæ, Lowe.

Genus 1. NUCULA, Lamarck.

Nucula nucleus, L. *Op. cit.* ii. p. 143, v. pl. 29. f. 2.

Common in the bay; brought in by the fishing-boats, and frequent in the stomachs of cod and haddocks.

Nucula nitida, G. B. Sowerby. *Op. cit.* ii. p. 140, v. pl. 29. f. 3 & 3 a.

Not rare off the East Rocks in sandy gravel between the rocky ridges, and in the stomachs of haddocks and cod.

K

Nucula tenuis, Mont. *Op. cit.* ii. p. 151, v. pl. 29. f. 4.
From deep water and the stomachs of cod and haddocks.

Genus 2. LEDA, Schumacher.

Leda minuta, Müller. *Op. cit.* ii. p. 155, v. pl. 29. f. 6.
Common in deep water and the stomachs of flounders.

Genus 4. PECTUNCULUS, Lamarck.

Pectunculus glycymeris, L. *Op. cit.* ii. p. 166, v. pl. 30. f. 2.
Abundant in the bay; generally brought in by the fishing-
boats.

Genus 5. ARCA, L.

Arca tetragona, Poli. *Op. cit.* ii. p. 180, v. pl. 30. f. 6.
Instead of the clusters in which it appears in the chinks of
the rocks in the Channel Islands, solitary examples only are
dredged off the bay in deep water.

Fam. 8. Kelliidæ, Forbes & Hanley.
Genus 2. MONTACUTA, Turton.

Montacuta bidentata, Mont. *Op. cit.* ii. p. 208, v. pl. 31. f. 8.
Abundant in shell-débris on the West Sands.

Montacuta ferruginosa, Mont. *Op. cit.* ii. p. 210, v. pl. 31. f. 9.
Common in the deep water of the bay and in the stomachs
of cod, haddocks, and flounders; also in the shell-débris on
the West Sands after storms.

Genus 3. LASÆA, Leach.

Lasæa rubra, Mont. *Op. cit.* ii. p. 219, v. pl. 32. f. 1.
Abundant amongst algæ, in crevices under stones in rock-
pools, and in the cavities of shells between tide-marks.

Genus 4. KELLIA, Turton.

Kellia suborbicularis, Mont. *Op. cit.* ii. p. 225, v. pl. 32. f. 2.
Common under stones in rock-pools, and in the cavities of
old limpet- and other shells.

Fam. 9. Lucinidæ, D'Orbigny.

Genus 2. LUCINA, Bruguière.

Lucina borealis, L. *Op. cit.* ii. p. 242, v. pl. 32. f. 7.
Frequently brought in by the fishing-boats, though the
majority of the specimens are imperfect (single valves).

Genus 3. AXINUS, J. Sowerby.

Axinus flexuosus, Mont. *Op. cit.* ii. p. 247, v. pl. 33. f. 1.
Single valves occasionally from the fishing-boats, and on
the West Sands after storms.

Fam. 10. Carditidæ, Gray.

Genus CYAMIUM, Philippi.

Cyamium minutum, Fabricius. *Op. cit.* ii. p. 260, v. pl. 33. f. 5.
Common in shell-débris on the West Sands.

Fam. 11. Cardiidæ, Broderip.

Genus CARDIUM, L.

Cardium echinatum, L. *Op. cit.* ii. p. 270, v. pl. 34. f. 2.
Very abundant on the West Sands after storms, and in the
débris of the fishing-boats.

Cardium fasciatum, Mont. *Op. cit.* ii. p. 281, v. pl. 35. f. 3.
Not uncommon on the West Sands after storms, and in the
stomachs of cod, haddocks, and flounders.

Cardium nodosum, Turton. *Op. cit.* ii. p. 283, v. pl. 35. f. 4.
Dead valves occasionally dredged off the East Rocks in
3 to 4 fathoms.

Cardium edule, L.   *Op. cit.* ii. p. 286, v. pl. 35. f. 5.

Abundant in the muddy sand at the mouth of the river Eden. Cockle-gathering forms the occupation of some of the fisherwomen.

Cardium norvegicum, Spengler.   *Op. cit.* ii. p. 294, v. pl. 35. f. 7.

Not uncommon ; generally brought by the fishermen from deep water.

Fam. 12. Cyprinidæ, Geinitz.

Genus 2. CYPRINA, Lamk.

Cyprina islandica, L.   *Op. cit.* ii. p. 304, v. pl. 36. f. 2.

Common in deep water, and thrown ashore after storms. Some have rows of small adherent pearls.

Genus 3. ASTARTE, J. Sowerby.

Astarte sulcata, Da Costa.   *Op. cit.* ii. p. 311, v. pl. 37. f. 1 & 2.

Frequently brought up by the deep-sea lines of the fishermen. Semifossil valves of *A. borealis* are also not uncommon.

Astarte compressa, Mont.   *Op. cit.* ii. p. 315, v. pl. 37. f. 3 & 4.

Frequent in deep water.

Genus 4. CIRCE, Schumacher.

Circe minima, Mont.   *Op. cit.* ii. p. 322, v. pl. 37. f. 6.

Not uncommon in deep water, and in the stomachs of cod, haddocks, and flounders.

Fam. 13. Veneridæ, Leach.

Genus 1. VENUS, L.

Venus exoleta, L.   *Op. cit.* ii. p. 327, v. pl. 38. f. 1.

Abundant in deep water, and on the beach after storms.

Venus lincta, Pulteney. *Op. cit.* ii. p. 330, v. pl. 38. f. 2.
Common in deep water, and thrown plentifully on the West
Sands after storms.

Venus fasciata, Da Costa. *Op. cit.* ii. p. 334, v. pl. 38. f. 4.
In 3 to 4 fathoms off the East Rocks, and from the deep-
sea lines of the fishermen. Dead valves are common amongst
the gravel at the East Rocks.

Venus casina, L. *Op. cit.* ii. p. 337, v. pl. 38. f. 5.
Occasionally procured in a perfect state from the deep-sea
lines of the fishermen. Single valves are abundant.

Venus ovata, Pennant. *Op. cit.* ii. p. 342, v. pl. 39. f. 1.
Common in deep water; generally procured from the fishing-
boats.

Venus gallina, L. *Op. cit.* ii. p. 344, v. pl. 39. f. 2 & 3.
Abundant on the West Sands after storms, and in a few
fathoms water on a sandy bottom all round.

Genus 2. TAPES, Mühlfeldt.

Tapes virgineus, L. *Op. cit.* ii. p. 352, v. pl. 39. f. 5.
Common in deep water and in the fishing-boats.

Tapes pullastra, Mont. *Op. cit.* ii. p. 355, v. pl. 39. f. 6.
Abundant between tide-marks amongst the muddy sand,
and occasionally in cavities bored by *Pholas crispata*.

Genus 3. LUCINOPSIS, Forbes & Hanley.

Lucinopsis undata, Pennant. *Op. cit.* ii. p. 363, v. pl. 40. f. 1.
Common on the sandy ground, and thrown in vast numbers
on the West Sands after storms.

Fam. 14. **Tellinidæ**, Latreille.

Genus 2. TELLINA, L.

Tellina crassa, Gmelin. *Op. cit.* ii. p. 373, v. pl. 40. f. 4.

Single valves of good size are not uncommon in the débris of the fishing-boats.

Tellina balthica, L. *Op. cit.* ii. p. 375, v. pl. 40. f. 5.

Abundant on the sandy beach at the mouths of the Eden and Tay, and on the West Sands after storms.

Tellina tenuis, Da Costa. *Op. cit.* ii. p. 379, v. pl. 41. f. 1.

Very common on the sandy ground everywhere; and dead valves occur on the West Sands throughout the year.

Tellina fabula, Gronovius. *Op. cit.* ii. p. 382, v. pl. 41. f. 2.

Only less common than the last species on the same ground.

Tellina pusilla, Philippi. *Op. cit.* ii. p. 388, v. pl. 41. f. 5.

Rather frequent in deep water, and in the stomachs of haddocks and flounders.

Genus 3. PSAMMOBIA, Lamarck.

Psammobia tellinella, Lamk. *Op. cit.* ii. p. 392, v. pl. 42. f. 1.

Worn valves occasionally found amongst the deep-sea lines of the fishermen.

Psammobia ferröensis, Chemnitz. *Op. cit.* ii. p. 396, v. p. 187, pl. 42. f. 3.

Abundant and in fine condition on a sandy bottom off the West Sands. Often thrown ashore in large numbers near the estuary of the Eden.

Genus 4. DONAX, L.

Donax vittatus, Da Costa. *Op. cit.* ii. p. 402, v. pl. 42. f. 5.

Very abundant on the West Sands after storms, and on sandy ground in a few fathoms.

Fam. 15. **Mactridæ**, Fleming.

Genus 2. MACTRA, L.

Mactra solida, L. *Op. cit.* ii. p. 415, v. pl. 43. f. 2.

Abundant on the sandy ground off the West Sands, and thrown in great numbers on the beach after storms.

Var. *elliptica* is common.

Mactra subtruncata, Da Costa. *Op. cit.* ii. p. 419, v. pl. 43. f. 3.

Equally common with the last species, and from the same ground.

Mactra stultorum, L. *Op. cit.* ii. p. 422, v. pl. 43. f. 4.

Very abundant on the same ground as the last two species. Var. *cinerea* is common.

Genus 3. LUTRARIA, Lamarck.

Lutraria elliptica, Lamk. *Op. cit.* ii. p. 428, v. pl. 44. f. 1.

Common on the West Sands after storms, and in muddy sand at the mouth of the river Eden.

Genus 4. SCROBICULARIA, Schumacher.

Scrobicularia prismatica, Mont. *Op. cit.* ii. p. 435, v. pl. 45. f. 1.

Not rare in deep water, on the West Sands after storms, and in the stomachs of cod and haddocks.

Scrobicularia alba, Müller. *Op. cit.* ii. p. 438, v. pl. 45. f. 3.

Less common than the foregoing, from the same ground, and in the stomachs of the same fishes.

Scrobicularia piperata, Bellonius. *Op. cit.* ii. p. 444, v. pl. 45. f. 5.

Common amongst the muddy sand at the mouth of the Tay, and frequently thrown on the West Sands after storms; also procured from the fishing-boats.

Fam. 16. **Solenidæ**, Latreille.

Genus 3. SOLEN, L.

Solen pellucidus, Pennant.   *Op. cit.* iii. p. 14, v. pl. 46. f. 4.

Common on the sandy ground, and thrown ashore in large numbers after storms ; occasionally in the stomachs of cod and haddocks.

Solen ensis, L.   *Op. cit.* iii. p. 16, v. pl. 47. f. 1.

Frequent on the sandy ground, and after storms on the West Sands.

Solen siliqua, L.   *Op. cit.* iii. p. 18, v. pl. 47. f. 2.

Abundant amongst the sand uncovered by low tides. The fishermen collect them for bait; and the children use them as scoops for digging in the sand.

Fam. 18. **Anatinidæ**, D'Orbigny.

Genus THRACIA, Leach.

Thracia papyracea, Poli.   *Op. cit.* iii. p. 36, v. pl. 48. f. 4 & 4 a.

Common on the sandy ground off the West Sands, and cast ashore plentifully after storms.

Fam. 19. **Corbulidæ**, Fleming.

Genus 3. CORBULA, Bruguière.

Corbula gibba, Olivi.   *Op. cit.* iii. p. 56, v. pl. 49. f. 6.

Off the East Rocks in a few fathoms, and on the beach after storms ; good specimens are also procured from the fishing-boats.

Fam. 20. **Myidæ**, Fleming.

Genus MYA, L.

Mya arenaria, L.   *Op. cit.* iii. p. 64, v. pl. 50. f. 1.

Frequent in the muddy sand at the mouth of the Eden. Distorted valves are common.

Mya truncata, L.   *Op. cit.* iii. p. 66, v. pl. 50. f. 2.

Abundant off the mouth of the Eden, and on the beach after storms.

Fam. 21. **Saxicavidæ**, Swainson.

Genus 2. SAXICAVA, Fleurian de Bellevue.

Saxicava rugosa, L.   *Op. cit.* iii. p. 81, v. pl. 51. f. 3 & 4.

Common at low-water mark amongst the rocks in crevices and holes in sandstone and shale, as well as inside empty limpet-shells and *Balani*. Often firmly adherent to its cavity by a byssus.

Fam. 23. **Pholadidæ**, Gray.

Genus 1. PHOLAS, Lister.

Pholas candida, L.   *Op. cit.* iii. p. 107, v. pl. 52. f. 2.

Rarely found in shale at the Castle Rocks; commonly met with on the beach after storms, sometimes in a perfect condition.

Pholas crispata, L.   *Op. cit.* iii. p. 112, v. pl. 53. f. 1.

Abundant in the soft shale and sandstone at East and West Rocks, and especially opposite the castle. Occasionally the siphons are observed protruding through sand which coats some of the ledges. Young specimens are often cast ashore on the West Sands in water-logged and decayed wood, whence they are extracted by the sea-fowl.

Genus 3. XYLOPHAGA, Turton.

Xylophaga dorsalis, Turton.   *Op. cit.* iii. p. 120, v. pl. 53. f. 4.

Not common; several living specimens occurred in the wood of a submerged thorn tree.

L

Order SOLENOCONCHIA.

Fam. Dentalidæ, H. & A. Adams.

Genus DENTALIUM, L.

Dentalium entalis, L.   *Op. cit.* iii. p. 191, v. pl. 55. f. 1.

Occurs on the West Sands in a living state after some storms. The specimens procured from the fishing-boats are generally tenanted by *Sipunculi.* Common.

Class GASTEROPODA.

Order I. CYCLOBRANCHIATA.

Fam. Chitonidæ, Guilding.

Genus CHITON, L.

Chiton fascicularis, L. Jeffreys, Brit. Moll. iii. p. 211,
v. pl. 55. f. 3.

Abundant under stones between tide-marks. This species,
like the limpet, forms considerable cavities in sandstone, so
that specimens become almost immersed.

Chiton cinereus, L. *Op. cit.* iii. p. 218, v. pl. 55. f. 2.

Common in deep water, and in the stomachs of the cod,
haddock, and flounder.

Chiton marginatus, Pennant. *Op. cit.* iii. p. 221, v. pl. 56. f. 5.

Frequent under stones between tide-marks.

Chiton ruber, Lowe. *Op. cit.* iii. p. 224, v. pl. 56. f. 4.

Occasionally between tide-marks at the East Rocks.

Chiton lævis (Pennant), Mont. *Op. cit.* iii. p. 226, v. pl. 56. f. 6.

Under stones at the verge of extreme low water during
spring tides. Rare; but the examples are large.

Order II. PECTINIBRANCHIATA.

Fam. 1. Patellidæ, Guilding.

Genus 1. PATELLA, Lister.

Patella vulgata, L., and var. *depressa*. *Op. cit.* iii. p. 236,
v. pl. 57.

Common everywhere; occasionally eaten. The soft shale
and sandstone are extensively pitted by this form.

Genus 2. HELCION, De Montfort.

Helcion pellucidum, L.   *Op. cit.* iii. p. 242, v. pl. 58. f. 1 & 2.
Abundant on the blades of the tangles; while var. *lævis* occurs in hollows at the bases of the stems.

Genus 3. TECTURA, Cuvier.

Tectura testudinalis, Müller.   *Op. cit.* iii. p. 246, v. pl. 58. f. 3.
Common under stones near low-water mark.

Tectura virginea, Müller.   *Op. cit.* iii. p. 248, v. pl. 58. f. 4.
Under stones near low-water mark; nearly as common as the foregoing.

Tectura fulva, Müller.   *Op. cit.* iii. p. 250, v. pl. 58. f. 5.
A single specimen on the West Sands after a storm.

Fam. 2. **Fissurellidæ**, Fleming.

Genus 2. EMARGINULA, Lamarck.

Emarginula fissura, L.   *Op. cit.* iii. p. 259, v. pl. 59. f. 2.
Not uncommon in deep water; and worn specimens amongst the shell-débris on sands.

Fam. 3. **Capulidæ**, Fleming.

Genus CAPULUS, De Montfort.

Capulus ungaricus, L.   *Op. cit.* iii. p. 269, v. pl. 59. f. 6.
Frequent in deep water, and often brought in by the fishing-boats.

Fam. 7. **Trochidæ**, D'Orbigny.

Genus 2. TROCHUS, Rondeletius.

Trochus tumidus, Mont.   *Op. cit.* iii. p. 307, v. pl. 62. f. 2.
Common on hard ground in the bay, in the stomach of the cod and haddock, and on the West Sands after storms.

Trochus cinerarius, L. *Op. cit.* iii. p. 309, v. pl. 62. f. 3.

In great abundance on stones and rocks between and beyond tide-marks.

Trochus zizyphinus, L. *Op. cit.* iii. p. 330, v. pl. 63. f. 6.

Not uncommon in deep water and on the beach after storms; rarely met with at extreme low water at the East Rocks. Var. *Lyonsii* is occasionally seen.

Fam. 9. Littorinidæ, Gray.

Genus 1. LACUNA, Turton.

Lacuna crassior, Mont. *Op. cit.* iii. p. 344, v. pl. 64. f. 2.

From tangle-roots and in old shells in the laminarian region, and from deep water; not rare.

Lacuna divaricata, Fab. *Op. cit.* iii. p. 346, v. pl. 64. f. 3.

On Fuci and laminarian blades at and beyond low-water mark. The colourless variety is not uncommon; and the same may be said of var. *quadrifasciata.*

Lacuna puteolus, Turton. *Op. cit.* iii. p. 348, v. pl. 64. f. 4.

With the former at low-water mark; less common than the foregoing.

Lacuna pallidula, Da Costa. *Op. cit.* iii. p. 351, v. pl. 64. f. 5 & 5 a.

On the West Sands after storms, and from the laminarian region.

Genus 2. LITTORINA, Férussac.

Littorina obtusata, L. *Op. cit.* iii. p. 356, v. pl. 65. f. 1 & 1 a.

Very common (with varieties) on stones and rocks between tide-marks.

Littorina rudis, Maton. *Op. cit.* iii. p. 364, v. pl. 65. f. 3, 3 a, & 3 b.

Abundant on the rocks near high-water mark.

Littorina litorea, L.   *Op. cit.* iii. p. 368, v. pl. 65. f. 5 & 5 *a*.
Between tide-marks in vast numbers. Often eaten.

Genus 3. RISSOA, Fréminville.

Rissoa parva, Da Costa.   *Op. cit.* iv. p. 23, v. pl. 67. f. 3 & 4.
In great numbers on the seaweeds in the laminarian region
all round, especially off the East Rocks. Var. *interrupta* is
also common in shell-sand.

Rissoa striata, Adams.   *Op. cit.* iv. p. 37, v. pl. 68. f. 2.
Common under stones between tide-marks. The var. *arctica*
is the prevailing form.

Rissoa soluta, Philippi.   *Op. cit.* iv. p. 45, v. pl. 68. f. 7.
Occasionally in deep water and in shell-sand.

Rissoa semistriata, Mont.   *Op. cit.* iv. p. 46, v. pl. 68. f. 8.
From deep water and in shell-sand; not common.

Genus 4. HYDROBIA, Hartmann.

Hydrobia ulvæ, Pennant.   *Op. cit.* iv. p. 52, v. pl. 69. f. 1.
In great abundance in the brackish pools near the mouth of
the river Eden.

Fam. 11. **Skeneidæ**, Clark.

Genus 1. SKENEA, Fleming.

Skenea planorbis, Fab.   *Op. cit.* iv. p. 65, v. pl. 70. f. 1.
Common in rock-pools amongst *Ceramium* and other algæ.

Genus 2. HOMALOGYRA, Jeffreys.

Homalogyra rota, Forbes & Hanley.   *Op. cit.* iv. p. 71,
v. pl. 70. f. 3.
In shell-débris from the West Sands. Dead.

Fam. 13. **Turritellidæ**, Clark.

Genus TURRITELLA, Lamarck.

Turritella terebra, L. *Op. cit.* iv. p. 80, v. pl. 70. f. 6.
Common in deep water. Var. *nivea* is also not rare. A favourite food of the codfish, probably in many cases on account of its tenant the hermit crab.

Fam. 15. **Scalariidæ**, Broderip.

Genus SCALARIA, Lamk.

Scalaria Trevelyana, Leach. *Op. cit.* iv. p. 93, v. pl. 71. f. 4.
From the fishing-boats, and on the West Sands after storms. Rather rare.

Fam. 16. **Pyramidellidæ**, Gray.

Genus 2. ODOSTOMIA, Fleming.

Odostomia rissoïdes, Hanley. *Op. cit.* iv. p. 122, v. pl. 73. f. 4.
Common in shell-sand.

Odostomia clathrata, Jeffreys. *Op. cit.* iv. p. 148, v. pl. 74. f. 9.
Worn specimens not uncommon in shell-débris from the West Sands.

Odostomia indistincta, Montagu. *Op. cit.* iv. p. 149.
v. pl. 75. f. 1.
Occasionally in shell-sand from the West Sands.

Odostomia interstincta, Montagu. *Op. cit.* iv. p. 151.
v. pl. 75. f. 2.
In shell-sand from the same locality.

Odostomia nitidissima, Montagu. *Op. cit.* iv. p. 173.
v. pl. 76. f. 8.
In débris from the West Sands.

Fam. 19. **Eulimidæ**, H. & A. Adams.

Genus EULIMA, Risso.

Eulima bilineata, Alder.   *Op. cit.* iv. p. 210, v. pl. 77. f. 8.
In muddy débris in old shells from deep water.

Fam. 20. **Naticidæ**, Swainson.

Genus NATICA, Adanson.

Natica islandica, Gmelin.   *Op. cit.* iv. p. 214, v. pl. 78. f. 1.
A single specimen from the fishing-boats.

Natica catena, Da Costa.   *Op. cit.* iv. p. 220, v. pl. 78. f. 4.
Common on the sandy bottom off the West Sands. The
ova occur abundantly in July and August.

Natica Alderi, Forbes.   *Op. cit.* iv. p. 224, v. pl. 78. f. 5.
Less abundant than the foregoing, on the same ground.

Fam. 22. **Velutinidæ**, Gray.

Genus 1. LAMELLARIA, Mont.

Lamellaria perspicua, L.   *Op. cit.* iv. p. 235, pl. 3. f. 6,
v. pl. 79. f. 2 & 2 a.

Common under stones between tide-marks, especially in
rock-pools. The figure of the living specimen in Dr. Gwyn
Jeffreys's work (iv. pl. 3. f. 6) was copied from a coloured
drawing of my sister's. The colours of this species vary in a
remarkable manner, as shown in Plate III. figs. 2–10.

Genus 2. VELUTINA, Fleming.

Velutina lævigata, Pennant.   *Op. cit.* iv. p. 240, v. pl. 79. f. 4.

Frequent in the laminarian region, and on the West Sands
after storms; occasionally under stones at the verge of low
water.

Fam. 23. **Cancellariidæ**, Forbes & Hanley.

Genus 2. TRICHOTROPIS, Broderip & Sowerby.

Trichotropis borealis, Brod. & Sowerb.   *Op. cit.* iv. p. 245,
v. pl. 79. f. 6.

A single specimen from the stomach of a cod.

Fam. 24. **Aporrhaidæ**, Troschel.

Genus APORRHAÏS, Da Costa.

Aporrhaïs pes-pelecani, L.   *Op. cit.* iv. p. 250, v. pl. 80. f. 1.

Abundant on the West Sands after storms, and frequent in the débris of the fishing-boats.

Fam. 27. **Buccinidæ**, Fleming.

Genus 1. PURPURA, Bruguière.

Purpura lapillus, L.   *Op. cit.* iv. p. 276, v. pl. 82. f. 1.

Very abundant between tide-marks on rocks and stones. Varieties are common.

Genus 2. BUCCINUM, L.

Buccinum undatum, L.   *Op. cit.* iv. p. 285, v. pl. 82. f. 2-5.

Common in the living state (var. *littoralis*) in pools at the East Rocks, especially where a stream of salt water rushes through the seaweeds. This species spawns in October, November, and the following months; the young are found in swarms on the egg-cases in May. Frequent on the West Sands after storms.

Genus 5. TROPHON, De Montfort.

Trophon truncatus, Ström.   *Op. cit.* iv. p. 319, v. pl. 84. f. 6.

Not uncommon in the fishing-boats, and the stomachs of the cod, haddock, and flounder.

M

Genus 6. FUSUS, Bruguière.

Fusus antiquus, L.   *Op. cit.* iv. p. 323, v. pl. 85. f. 1 & 2.

Abundant in the coralline zone, and frequently thrown on shore after storms.

Fusus gracilis, Da Costa.   *Op. cit.* iv. p. 335, v. pl. 86. f. 2.

Common on the West Sands after storms, and often brought in by the fishing-boats.

Fusus propinquus, Alder.   *Op. cit.* iv. p. 338, v. p. 219, pl. 86. f. 3.

Occasionally procured from the deep-sea lines of the fishermen.

Fam. 29. **Nassidæ**, Stimpson.

Genus 1. NASSA, Lamk.

Nassa incrassata, Ström.   *Op. cit.* iv. p. 351, v. pl. 88. f. 1.

Common in the laminarian region and under stones between tide-marks; while worn shells are abundant in débris at the East Rocks. An egg-capsule is shown in Plate IX. fig. 4.

Fam. 30. **Pleurotomidæ**, Lovén.

Genus 1. DEFRANCIA, Millet.

Defrancia linearis, Mont.   *Op. cit.* iv. p. 368, v. pl. 89. f. 2.

From deep water; rather rare.

Genus 2. PLEUROTOMA, Lamk.

Pleurotoma costata, Donovan.   *Op. cit.* iv. p. 379, v. pl. 90. f. 3.

Occasionally in shell-débris from the West Sands.

Pleurotoma rufa, Montagu.   *Op. cit.* iv. p. 392, v. pl. 91. f. 6.

Common on the West Sands after storms.

Pleurotoma turricula, Montagu.   *Op. cit.* iv. p. 395, v. pl. 91. f. 7.

Abundant under the same circumstances.

Pleurotoma Trevelyana, Turton.   *Op. cit.* iv. p. 398,
v. pl. 91. f. 8.

Not uncommon in the stomachs of cod and haddock.

Fam. 31. **Cypræidæ**, Fleming.

Genus 2. CYPRÆA, L.

Cyprœa europœa, Mont.   *Op. cit.* iv. p. 403, pl. 7. f. 4,
v. pl. 92. f. 2.

Not uncommon at extreme low water, under stones in
pools in the same region at the East Rocks, and generally in
the laminarian region. The living animal represented in Dr.
Gwyn Jeffreys's work is from a coloured drawing by my sister.

Order IV. PLEUROBRANCHIATA, Gray.

Fam. 1. **Bullidæ**, Clark.

Genus 1. CYLICHNA, Lovén.

Cylichna cylindracea, Pennant.   *Op. cit.* iv. p. 415,
v. pl. 93. f. 4.

Abundant in deep water; generally thrown on the West
Sands after storms.

Genus 2. UTRICULUS, Brown.

Utriculus truncatulus, Bruguière.   *Op. cit.* iv. p. 421,
v. pl. 94. f. 2.

Occasionally from deep water, and in débris on sands.

Utriculus obtusus, Mont.   *Op. cit.* iv. p. 423, v. pl. 94. f. 3.

Not uncommon in deep water. A variety with a more extended spire is often met with in shell-débris from the West
Sands.

Genus 4. ACTÆON, De Montfort.

Actœon tornatilis, L.   *Op. cit.* iv. p. 433, v. pl. 95. f. 2.

Frequent off the West Sands, and thrown ashore after
storms.

M 2

Genus 6. SCAPHANDER, De Montfort.

Scaphander lignarius, L. *Op. cit.* iv. p. 443, v. pl. 95. f. 5.

Not uncommon in deep water, and thrown ashore alive after storms.

Genus 7. PHILINE, Ascanius.

Philine scabra, Müller. *Op. cit.* iv. p. 447, v. pl. 96. f. 1.

Abundant in deep water, and in the stomachs of cod, haddock, and flounders. The shells occur on the West Sands after storms.

Philine pruinosa, Clark. *Op. cit.* iv. p. 454, v. pl. 96. f. 6.

From the stomach of a haddock; rare.

Philine aperta, L. *Op. cit.* iv. p. 457, v. pl. 96. f. 8.

Occasionally from deep water; shells thrown on the West Sands after storms.

Fam. 2. **Aplysiidæ**, D'Orbigny.

Genus APLYSIA, L.

Aplysia punctata, Cuvier. *Op. cit.* v. p. 5, pl. 97. f. 1.

A single specimen (Plate III. fig. 1) was found in autumn amongst the sea-weeds of a large pool at extreme low water. No spots or other markings were present on the dull olive hue of the body.

Order V. NUDIBRANCHIATA, Cuvier.

Suborder 1. **PELLIBRANCHIATA.**

Fam. 1. **Limapontiidæ**, Alder & Hancock.

Genus 1. LIMAPONTIA, Johnston.

Limapontia nigra, Johnston. *Op. cit.* v. p. 28, pl. 1. f. 5.

Not uncommon amongst the seaweeds under stones in rock-pools.

Suborder II. **POLYBRANCHIATA.**

Fam. 3. **Eolididæ,** D'Orbigny.

Genus 2. Eolis, Cuvier.

Eolis papillosa, L.; Alder & Hancock, Brit. Nud. Moll.
fam. 3, pl. 9.

Abundant in early spring (February) amongst the rocks
near low water, and occasionally at other times.

Eolis coronata, Forbes; A. & H. B. Nud. M. fam. 3, pl. 12.

Common in February and the spring months in the same
localities.

Eolis rufibranchialis, Johnston; A. & H. B. Nud. M.
fam. 3, pl. 16.

With the foregoing; common.

Eolis olivacea, Alder & Hancock; A. & H. B. Nud. M.
fam. 3, pl. 26.

Not uncommon under stones in pools at all seasons.

Eolis viridis, Forbes; A. & H. B. Nud. M. fam. 3,
pl. 32.

Abounds on the small hydroid zoophytes under stones in
rock-pools.

Eolis Farrani, Alder & Hancock; A. & H. B. Nud. M.
fam. 3, pl. 35.

Occasionally occurs under stones at low-water mark at the
East Rocks. The fine purple variety has been found more
than once (Plate II. figs. 12 & 13). A specimen shows the
abnormality of a clavate median process between the oral
and dorsal tentacles.

Eolis Adelaidœ, Thompson; Jeffreys, Brit. Moll. v. p. 55.
A single specimen (Plate II. fig. 11) was found in a sandy pool in August.

Eolis exigua, Alder & Hancock; A. & H. B. Nud. M. fam. 3, pl. 37.
Not uncommon on laminarian blades thrown on the West Sands after storms.

Fam. 5. Dotonidæ.

Genus Doto, Oken.

Doto fragilis, Forbes; A. & H. B. Nud. M. fam. 3, pl. 5.
Occasionally on zoophytes brought in by the fishing-boats.

Doto coronata, Gmelin; A. & H. B. Nud. M. fam. 3, pl. 6.
Common in the débris of the fishing-boats ; small specimens are frequently found in groups on *Sertularia pumila* under stones in rock-pools near low-water mark; also on laminarian blades after storms. One example has an abnormal left tentacle (Plate II. fig. 14).

Fam. 6. Dendronotidæ.

Genus 1. DENDRONOTUS, Alder & Hancock.

Dendronotus arborescens, Müller; A. & H. B. Nud. M. fam. 3, pl. 3.
Not uncommon amongst zoophytes from deep water, on laminarian blades thrown on the West Sands, and occasionally at extreme low water. A white variety is sometimes seen.

Fam. 8. Tritoniidæ, H. & A. Adams.

Genus TRITONIA, Cuvier.

Tritonia Hombergi, Cuv.; A. & H. B. Nud. M. fam. 2, pl. 2.
From deep water. Both pale and deep reddish-brown

varieties are found on the zoophytes in the fishing-boats. Occasionally in the stomach of the cod.

Tritonia plebeia, Johnston; A. & H. B. Nud. M. fam. 2, pl. 3.

In vast numbers amongst the zoophytes from deep water, in the crevices of *Alcyonium digitatum* and on *Halecium* tossed ashore after storms. One specimen showed the abnormality of a bifid tail (Plate II. fig. 15).

Suborder III. ACANTHOBRANCHIATA.

Fam. 1. Polyceridæ.

Genus 1. ÆGIRUS, Lovén.

Ægirus punctilucens, D'Orbigny; A. & H. B. Nud. M. fam. 1, pl. 21.

Not uncommon under stones in rock-pools between tide-marks at the East Rocks.

Genus 2. TRIOPA, Johnston.

Triopa claviger, Müller; A. & H. B. Nud. M. fam. 1, pl. 20.

Fine specimens are occasionally found under stones near low-water mark. The same *Ergasilus* is parasitic on this as on *Doris.*

Genus 5. POLYCERA, Cuvier.

Polycera quadrilineata, Müller; A. & H. B. Nud. M. fam. 1, pl. 22.

Not uncommon near low-water mark and in the laminarian region.

Polycera ocellata, Alder & Hancock; A. & H. B. Nud. M. fam. 1, pl. 23.

Gregarious under stones between tide-marks, but not common. They chiefly occur at the West Rocks.

Polycera Lessoni, D'Orbigny ; A. & H. B. Nud. M. fam. 1,
pl. 24.

On a laminarian blade after an October storm ; one specimen.

Genus 6. ANCULA, Lovén.

Ancula cristata, Alder ; A. & H. B. Nud. M. fam. 1,
pl. 25.

Not rare under stones in rock-pools, chiefly at the East
Rocks.

Genus 8. GONIODORIS, Forbes.

Goniodoris nodosa, Mont. ; A. & H. B. Nud. M. fam. 1,
pl. 18.

Abundant between tide-marks under stones in rock-pools
and elsewhere, throughout the year.

Fam. 2. **Dorididæ.**

Genus DORIS, L.

Doris tuberculata, Cuvier ; A. & H. B. Nud. M. fam. 1, pl. 3.

Abundant under rocky ledges and under stones in pools.
Its parasite, *Ergasilus,* is common.

Doris Johnstoni, Alder & Hancock ; A. & H. B. Nud. M.
fam. 1, pl. 5.

[Plate II. fig. 16.]

Occasionally under stones in pools between tide-marks
at the East Rocks. The same *Ergasilus* occurs on this
species.

Doris repanda, Alder & Hancock ; A. & H. B. Nud. M.
fam. 1, pl. 6.

Abundant at all seasons amongst the rocks between tide-
marks. A specimen in the act of spawning is represented

in Plate VII. figs. 13 & 14, and the ova after deposition in fig. 15.

Doris aspera, Alder & Hancock; A. & H. B. Nud. M. fam. 1. pl. 9. f. 1–9.

Common under stones in pools near low-water mark.

Doris bilamellata, L.; A. & H. B. Nud. M. fam. 1, pl. 11.

Abundant between tide-marks at all times; in swarms in March.

Doris pilosa, Müller; A. & H. B. Nud. M. fam. 1, pl. 15.

Common at all seasons between tide-marks.

Class CEPHALOPODA.

Order DIBRANCHIATA, Owen.

Fam. 1. **Teuthidæ**, Owen.

Genus 1. OMMATOSTREPHES, D'Orbigny.

Ommatostrephes todarus, Delle Chiaje, Mem. An. s. Vert. Nap. iv. Mem. ii. tav. 60.

Frequently thrown in numbers on the West Sands, especially after April storms.

Genus 2. LOLIGO, Schneider.

Loligo vulgaris, Lamarck. Jeffreys, Brit. Moll. v. p. 130, pl. 5. f. 2.

The shells are sometimes found in the stomachs of codfish. The spawn of this species is frequent.

Genus 4. SEPIOLA, Rondelet.

Sepiola Rondeletii, Leach. *Op. cit.* v. p. 136, pl. 6. f. 2.

The sole specimen occurred in the stomach of a flounder.

N

Fam. 3. Octopidæ, D'Orbigny.

Genus 2. ELEDONE, Leach.

Eledone cirrosa, Lamk. *Op. cit.* v. p. 146, pl. 7. f. 2.

Occasionally in pools between tide-marks, and on the West Sands after storms; common in the stomachs of cod and haddock.

Subkingdom *ANNULOSA.*

Series I. ANNULOIDA.

Class ECHINODERMATA.

The Echinoderms of St. Andrews, though plentiful, are by no means remarkable, being those generally distributed over the north-east coast. We do not find the rosy feather, the bird's-foot, and the little cushion starfishes so abundant on the southern and western shores, the former extending to the tangles of Shetland and far into the Atlantic. The beautiful pale bluish-purple *Asterias glacialis,* so common under littoral stones at Herm, and the great *Luidia Savignii* of the surrounding currents are absent (though the former occasionally occurs on the east coast of Scotland); and so is *Asterias Mülleri* of the Hebridean lochs. The northern waters are further distinguished by the piper (*Cidaris papillata*) and swarms of *Echinus norvegicus*; and the southern by the splendid condition of the purple, Fleming's, and the silky-spined urchins. The profusion of sea-cucumbers characteristic of certain parts affords another contrast: thus, as truly said by Prof. Edward Forbes, the giant of the race seems to have rallied all his subjects around him in the rich tangle-forests of the Zetlandic voes. The vast numbers of *Synapta tenera* on the muddy banks of the numerous islets in the Sound of Harris is distinctive, just as the abundance of *Synapta Gallienii* (which the Rev. Mr. Norman seems inclined to link on to *S. inhærens*) is in Belgrave Bay, Guernsey, and a large brownish-purple species on the south-west coast of Ireland.

The places of the rare are filled by a multitude of the common forms, which abound on the beach after storms, and under stones between tide-marks, or are dredged in the surrounding waters. The ease with which the development of

N 2

the young of this group can be observed opens up an excellent field for future investigators.

I have to thank the Rev. A. M. Norman for his kind assistance in revising the following list, and determining several Holothuroidea.

Order II. OPHIUROIDEA.

Fam. 2. Ophiuridæ.

Genus 4. OPHIOTHRIX, Müller & Troschel.

Ophiothrix fragilis, O. F. Müller ; Rev. A. M. Norman, Ann. & Mag. Nat. Hist. February 1865, p. 107.

Abundant under stones in rock-pools and near low-water mark, and dredged in the water beyond to a considerable depth. Many of the stones in the pools are covered with the ova of this species about the middle of November ; and some of the starfishes have them attached to the disk.

Genus 5. AMPHIURA, Forbes.

Amphiura filiformis, O. F. Müller ; Norman, *op. cit.* p. 107.

Occasionally in the stomachs of haddocks.   Rare.

Amphiura Chiajii, Forbes ; Norman, *op. cit.* p. 107.

Vast numbers are thrown ashore on the West Sands after storms.   It is also common in the stomachs of the cod and haddock.

Amphiura elegans, Leach ; Norman, *op. cit.* p. 109.

Frequent under stones in rock-pools and near low water, especially towards the Rock and Spindle.

Genus 7. OPHIOCOMA, Agassiz.

Ophiocoma nigra, O. F. Müller ; Norman, *op. cit.* p. 111.

Not uncommon from deep water (by dredging and the deep-sea lines of the fishermen).   It does not occur in profusion, as in many parts of the Zetlandic and southern portions of our seas.

Genus 8. OPHIOPHOLIS, Müller & Troschel.

Ophiopholis aculeata, O. F. Müller; Norman, *op. cit.* p. 112.

Rather plentiful in deep water, and common in the stomach of the cod; occasionally under stones near low water at the East Rocks.

Genus 9. OPHIURA, Lamarck.

Ophiura lacertosa, Pennant; Norman, *op. cit.* p. 112.

Abundant off the West Sands, and thrown on the beach in great numbers after storms; it is then much preyed on by gulls.

Ophiura albida, Forbes; Norman, *op. cit.* p. 113.

Dredged off the East Rocks on a sandy bottom, and procured from the stomachs of haddocks.

Order III. ASTEROIDEA.

Fam. 1. **Astropectinidæ.**

Genus 10. ASTROPECTEN, Linck.

Astropecten irregularis, Pennant; Norman, *op. cit.* p. 116.

Very abundant on the West Sands after storms.

Genus 11. LUIDIA, Forbes.

Luidia Sarsii, Düben & Koren; Norman, *op. cit.* p. 118.

Occasionally from the deep-sea lines of the fishermen. It takes the place of the larger *L. Savignii* of the prolific waters of the Channel Islands.

Fam. 2. **Solastridæ.**

Genus 15. SOLASTER, Forbes.

Solaster papposus, L.; Norman, *op. cit.* p. 122.

Abundant on the West Sands after storms, and at all times at low water amongst the rocks.

Solaster endeca, L.; Norman, *op. cit.* p. 122.

Not uncommon on the West Sands after storms, but much less abundant than the foregoing.

Genus 18. CRIBRELLA, Agassiz.

Cribrella sanguinolenta, O. F. Müller; Norman, *op. cit.* p. 124.

Very common between tide-marks, often hanging to the dripping sides and roofs of caverns. A large and much softer variety occasionally occurs. The greater diameter in several instances reaches 5 inches; and one exceeds this size.

Fam. 3. Asteriadæ.

Genus 20. ASTERIAS, L.

Asterias rubens, L.; Norman, *op. cit.* p. 128.

Abundant between tide-marks and beyond. Many singular varieties, from the loss or partial reproduction of the rays, occur, the most remarkable, perhaps, being that represented in Plate VI. fig. 1. A specimen shows five large rays, two of which are formed by the splitting of one arm, while in the interspace are two small rays situated one over the other. They spawn in November; and many are found in the peculiar stool-like position, grasping the ova, at this season. The same posture, however, is sometimes assumed when devouring *Littorina obtusata* or other mollusks. Examples with developing arms are shown in Plate VI. figs. 2 & 3.

Asterias violacea, O. F. Müller; Norman, *op. cit.* p. 128.

As common as the foregoing, and even more so between tide-marks.

Asterias hispida, Pennant; Norman, *op. cit.* p. 128.

This species has only been seen at St. Andrews by Prof. Edward Forbes, who found several specimens on the sands after a storm in 1839. Although hundreds of small forms have been examined, no specific character has occurred to separate them from the foregoing (*A. rubens* and *A. violacea*).

Order IV. Echinoidea.

Fam. 1. Cidaridæ.

Genus Echinus, L.

Echinus esculentus, L.; Forbes, Brit. Starfishes, p. 149.

Common amongst the tangles at extreme low water, and in the laminarian region; young specimens occur under stones between tide-marks. In many the intestinal canal is loaded with fragments of laminarian stalks, pieces of *Delesseria*, and other seaweeds covered with *Membranipora*; in some there are fragments of the shells of *Balani* and tubes of *Serpulæ*.

Echinus miliaris, Leske; Forbes, Brit. Starf. p. 161.

Not uncommon under stones in rock-pools.

Echinus Flemingii, Ball; Forbes, Brit. Starf. p. 164.

Occasionally in deep water off the bay, and thrown on the West Sands after storms. The specimens are much less than those of the Channel Islands.

Genus Toxopneustes, Agassiz.

Toxopneustes dröbachiensis, O. F. Müller; Forbes, Brit. Starf. p. 172 (as *Echinus neglectus*).

Not uncommon on the West Sands after storms. The specimens are smaller than those from the Channel Islands.

Fam. 2. Clypeastridæ.

Genus Echinocyamus, Leske.

Echinocyamus angulosus, Leske; Forbes, Brit. Starf. p. 175.

Abundant in deep water and in the stomachs of the cod, flounder, and haddock. Worn specimens also occur at the East Rocks amongst the shell-gravel.

Fam. 3. **Spatangidæ.**

Genus SPATANGUS, Klein.

Spatangus purpureus, O. F. Müller; Forbes, Brit. Starf.
p. 182.

Not uncommon in deep water, and occasionally thrown on
the West Sands by storms.

Genus ECHINOCARDIUM, Gray.

Echinocardium cordatum, Pennant; Forbes, Brit. Starf.
p. 190.

Very common off the West Sands, and tossed on the beach
at all seasons.

Echinocardium ovatum, Leske; Forbes, Brit. Starf. p. 194
(as *Amphidotus roseus*).

Occurs in deep water, and on the beach after storms; somewhat rare.

Order V. HOLOTHUROIDEA.

Fam. **Psolidæ.**

Genus PSOLUS, Oken.

Psolus phantapus, L.; Forbes, Brit. Starf. p. 203.

Occasionally from deep water, and brought in by the fishing-
boats.

Fam. **Pentactæ.**

Genus CUCUMARIA, Blainville.

Cucumaria ——?

A large purplish-brown species, common on the West Sands
after storms. Mr. Norman thinks "this is probably the
species found by Mr. Goodsir off the Fifeshire coast, and referred to *C. frondosa* by E. Forbes. It is very like that species

in most of its characters, especially in the total absence of
skin-spicules, and in the form of the tentacular spicula, which
are elongated and cribrose. It appears to differ from *C. frondosa* in its very thick teat, and especially in appearing to have
feet scattered over the body between the regular rows. At
the same time it is possible that the firmness may be due to
a state of rigid contraction from having been beaten about in
a storm when alive; and with respect to the latter, the pores
may not mark contracted feet. . . . . It does not correspond badly
with the description of *C. Drummondi*, a species unknown to
me."

Cucumaria elongata, Düben & Koren; Norman, Zetlandic
    Fauna, Rep. Brit. Assoc. 1868, p. 316.

= *Cucumaria pentactes*, Forbes (partim), the centre figure in woodcut
    p. 213.

Specimens are occasionally brought from the coralline
ground by the fishermen.

Cucumaria Hyndmanni, Thompson; Forbes, Brit. Starf.
    p. 225.

Not uncommon in the stomachs of haddocks and cod.

Cucumaria lactea, Forbes & Goodsir; Forbes, Brit. Starf.
    p. 231.

[Plate IV. fig. 5, and Plate IX. fig. 5.]

Abundant in the coralline region amongst zoophytes.
Young specimens are numerous in June.

Genus THYONE, Oken.

Thyone fusus, O. F. Müller; Forbes, Brit. Starf. p. 233.
Common in the stomachs of cod and haddock.

Genus THYONIDIUM, Düb. & Koren.

Thyonidium Dubeni, Norman, *op. cit.* p. 317.

Occasionally in the stomachs of the cod and haddock. Mr.
Norman states that he has found it on the coast of Ireland, as

well as in Shetland. He observes (*in lit.*) that in this form
there are no skin-spicula; feet with a large, circular, cribrose
plate at the end, no spicula on sides; tentacles cased in large
cribrose spicula of varied form—elongated, short, or most
elegantly irregular and branched.

Thyonidium commune, Forbes & Goodsir; Forbes, Brit.
Starf. p. 217, and Norman, *op. cit.* p. 317.

A fragmentary specimen in the stomach of a cod.

Fam. **Synaptidæ.**

Genus SYNAPTA, Eschsch.

Synapta inhærens, O. F. Müller; Dr. Herapath, Journ.
Micr. Sc. 1865, p. 4.

[Plate IV. fig. 4, and Plate IX. figs. 6, 7, & 8.]

The typical form occurs between tide-marks, as well as in
the laminarian region, the anchor-plates having six apertures
surrounding the central, and comparatively few openings in

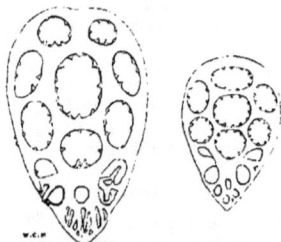

the narrow part to which the anchor is attached (see smaller
figure in woodcut, which represents both forms × 210 diam.).
Such agrees closely with examples from the Channel Islands,
the Hebrides, and other parts. An imperfect specimen from
the stomach of a haddock diverges very considerably in the
form of its anchor-plates (woodcut, larger figure), since the

whole plate is much larger, and there are generally seven apertures round the central, instead of six as in the former case ; while the slits in the smaller end (to which the anchor is attached) are much more numerous and linear. Various abnormal anchors occur in *S. inhærens*, such as one with five flukes (a bifid process on the summit, a bifid fluke and a normal serrated fluke), or an anchor with several processes on the stalk.

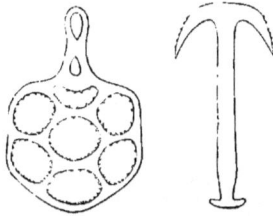

Plate and anchor of *Synapta tenera*, Norman, from Lochmaddy.

Class ANNELIDA.

The marine annelids have sometimes been considered an un-inviting group, dimly associated with parasites and earthworms. In regard, however, to beauty of form and colour, wonderful structure and habits, they are not surpassed by any invertebrate class. The splendid bristles of the Aphroditidæ, constantly glistening with all the hues of a permanent rainbow, the bril-liant colours of the Phyllodocidæ, Hesionidæ, and Nereidæ, and the gorgeous branchial plumes of the Terebellidæ, the Sabel-lidæ, and the Serpulidæ can only be compared with the most beautiful types of butterflies and birds. The structures formed by many exhibit an amount of precision and skill equal to that of the most remarkable insects. Thus, at St. Andrews, the common *Pectinaria belgica* fashions a tube like a straight horn of minute pebbles, carefully selected and admirably fixed to each other by a whitish cement. In the placing of these together there is no haphazard, but angle fits angle as in a skilfully built wall, and no profusion of the whitish cement hides slovenly masonry. There is much similarity in the ordinary tubes; dozens may be examined without observing any noteworthy structural difference. All have the same blending of the white or light-coloured grains with the yellow, the brown, and the black. There is no chance grouping, so as to cause the tube to be out of harmony with its surroundings; but the whole tone is such that it can with difficulty be distinguished from the sand. Some annelids, again, secrete transparent tubes of the aspect and toughness of crow-quills; while others cement the mud into caoutchouc-like pipes, fix gravel, stones, and shells by the same means into convenient tunnels, or rely on the parchment-like tenacity of a tube formed solely of one or more layers of their re-markable secretion. The interest in the group is further heightened by the brilliant phosphorescence characteristic of many, and the powers which others have of perforating sand, limestone, shells, aluminous shale, sandstone, and other rocks.

The annelids are not devoid of interest even in an economical point of view. All round the coasts of Britain the *Arenicola marina* (common lobworm) is generally used as bait, and here and there *Nephthys* and *Nereilepas fucata*. On the prolific shores of the Channel Islands the great abundance of the Nereidæ is of considerable importance to the inhabitants, since two of the most plentiful (viz. *Nereis cultrifera*, Grube, and *N. diversicolor*, Müller) are extensively used in fishing. The fishermen constantly search for them with a pointed instrument resembling a spear (see annexed woodcut), and keep them in vessels amongst a little sand and seaweed. They are much employed in catching whiting, the latter, again, being used as bait in conger-fishing. In the same islands one of the most esteemed baits is the large *Marphysa sanguinea*, which reaches the length of two feet. It is termed "varme" by the fishermen, and is highly prized both for the capture of ordinary white fish and dogfish. The annelids are kept alive in vessels amongst seaweed—or rather the anterior segments only, no more than three or four inches of this region being retained, since experience has shown that, unless so treated, the animals will break off posterior fragments, which, putrefying, soon cause the death of the whole. The natives of the Fiji group much esteem a form allied to the British *Lysidice ninetta* as an article of diet, and they predict its annual appearance in their seas with unerring precision by observing the phases of the moon, as at Samoa. It is called "Palolo" by the Samoans and Tongese, and "Mbalolo," Dr. Denis Macdonald informs us [*], by the Fijians. This annelid occurs in numbers so vast that it is collected by the natives as a dainty and nutritious food; and it is so much prized that formal presents of it are often sent considerable distances from certain chiefs to others.

* Linn. Trans. vol. xxii. p. 237 (1859).

whose small dominions do not happen to be visited by the
Palolo.   Dr. Macdonald thinks the tendency to transverse
fission exhibited by the annelids (since they are seldom got
entire) may be connected with the diffusion of the ova, and
not with the development of new forms—a conclusion the more
likely, though by no means necessary.   He states that the
species had been supposed to exhibit an alliance with *Arenicola*, but that the anatomical characters refer it to the Nereidæ.
As already mentioned, it ought rather to be classed with the
Eunicidæ.   If the Palolo has similar habits to the *Lysidice*
of our southern coasts (that is, dwells in fissures and crevices
of the rocks at and near low water), it probably leaves its
retreats for the purpose of depositing ova.   Lastly, *Echiurus*
is used as bait by the Belgian fishermen; and a *Sipunculus* is
employed as food by the Chinese, whose varied taste ranges
from trepangs to edible birds' nests.

If the uses of the majority of the annelids are restricted in
the case of man, a very different condition holds with regard
to marine animals.   An examination of the stomachs of our
most valuable fishes shows how acceptable and important
a part they play in the supply of nutriment.   The large
number of species which a few hours' fishing on a rich coast
will produce with bait of *Nereis cultrifera* is strongly corro-
borative; indeed I should be inclined to place them even
before crabs and mollusks in respect of the avidity with which
fishes devour them.   The majority of the annelids of St.
Andrews are found in the stomachs of cod, haddock, whiting,
flounders, and other common fishes; and it is often puzzling
to explain how those which dwell in tubes under stones, in
fissures of rocks, and in other remote places have been obtained.   To give a satisfactory account of the food furnished
by this class to fishes would require an enumeration of every
family, and most of the genera and species, found in this
country; indeed I do not know a single form that would be
rejected.   It will suffice, on the present occasion, to notice a
few of the more conspicuous at St. Andrews.   The stomachs
of cod and haddock are frequently filled with sea-mice and
Polynoidæ; and another very common form is *Sigalion
Mathildæ*.   The Nereidæ (from the gigantic *Alitta virens*,

Sars, which often distends the stomachs of large cod, to the
smaller *Nereis pelagica* and *N. cultrifera* are universally
eaten. The somewhat uninviting *Trophonia plumosa* sometimes forms the sole food in the stomachs of large haddocks,
many hundreds occurring in a single fish. *Owenia filiformis*,
with its gravelly tubes, is a favourite diet of the same fish,
and of cod and flounders. The Terebellidæ and their sandy
tubes are also largely devoured; and even Serpulidæ are not
passed by. Moreover, in their young or larval forms they
constitute an important element in the food of the herring and
other fishes that feed near the surface of the water.

Many of the annelids of St. Andrews are common to the
whole British area; but some have not yet been found in other
parts of our seas : as this, however, is probably due to a larger
amount of attention having been directed to the locality, we
shall not at present particularize.

The fauna at St. Andrews is distinguished, as far as our
present knowledge extends, from the Zetlandic by the absence
of such striking forms as *Lætmonice*, *Panthalis*, *Nothria
conchylega*, *Terebella nebulosa*, *Pista cristata*, *Trichobranchus
glacialis*, and *Ditrypa arietina*; from that of the western
regions by the absence of *Spinther*, *Lepidonotus clava*, *Polynoë scolopendrina*, *Ophiodromus vittatus*, *Guttiola spectabilis*,
Terebella nebulosa, and *Pista cristata*; and of the southern
types we miss *Euphrosyne*, *Hermione*, *Polynoë areolata*, *Nereis
Marionii*, *Lysidice ninetta*, the Eunicidæ, the abundance
of the Chætopteridæ, *Sabellaria alveolata*, *Lepræa terstris*,
Sabella saxicava, *Protula*, and *Filigrana* between tide-marks.
The great preponderance of *Polynoë floccosa* in the south is
also an interesting feature.

Amongst the annelids that, besides other very common
forms, abound at St. Andrews, and therefore most characteristic
of it, are *Sigalion Mathildæ*, *Sthenelais limicola*, *Phyllodoce
laminosa*, *P. groenlandica*, *Nereis cultrifera*, *Alitta virens*, *Autolytus pictus*, *Aricia Cuvieri*, *Ophelia limacina*, *Trophonia plumosa*, *Nerine foliosa*, *Polydora ciliata*, *Capitella capitata*,
Sabellaria spinulosa, *Pectinaria belgica*, *Lanice conchilega*,
Sabella pavonia, and *Branchiomma vesiculosum*.

Some of the phosphorescent forms at St. Andrews have

already been noticed * ; so that in the mean time the remarks shall be confined to the Polynoidæ, three common species of which afford ready means of experiment. In *Harmothoë imbricata* irritation causes a series of bluish-green flashes at the points of attachment of the scales, and then a steady light for some time. Very pale specimens seem more irritable than ordinary forms. No pulsations of light are observed on the phosphorescent surface of the detached scales. On the whole the light in this species is characterized by its steadiness. It does not readily emit its phosphorescence when a little sulphuric ether is added to the water; nor does mechanical irritation in these circumstances cause any change in its manifestation. Acetic acid acts as a poison, causing a momentary gleam as the scales are thrown off, which wholly disappears with the death of the animal and the ejection of the proboscis. If strong methylated spirit be gradually added to the sea-water (in a small vessel), there is seldom phosphorescence if no mechanical irritation occurs; the animal perishes with all the scales on its back. The luminous emissions are similar when spirit is applied to the annelid in the air. *Polynoë floccosa* seems to be more irritable, and to emit its phosphorescence more readily than the foregoing at the same points. When one of the scales is detached, the greenish light is given off as if in pulsations from the surface of attachment, somewhat quickly at first, then slower, and finally disappearing. In *Evarne impar*, again, the detached scales give off a flashing light, such as might be caused by a swift series of waves, and which quite differs in character from that in *P. floccosa*.

The Gephyreans are not so abundant as on the muddy flats of the west and south, where swarms of the common forms are found in a single spadeful. The highly characteristic *Echiurus vulgaris*, however, occurs, often in great numbers; and though *Priapulus caudatus* is not met with in the littoral region, nor so large as in the Hebrides, still it is not rare in deep water, and is frequent in the stomachs of fishes.

* Ann. & Mag. Nat. Hist. 4th ser. 1872, vol. ix. pp. 6 & 7.

The Nemerteans, again, are especially abundant between tide-marks, though some range thence to deep water, and a few occur only in the latter. *Amphiporus lactifloreus* is common under stones, and *Lineus gesserensis* and *Cephalothrix linearis* in still greater numbers, especially in muddy places. The great *Lineus marinus* is frequently found under stones, and occasionally in the pools. Near low water the *Tetrastemma* (such as *T. melanocephala, T. candida, T. vermicula, T. flavida*, and occasionally *T. dorsalis*) occur in varying numbers, the latter, however, attaining its maximum amongst the red seaweeds in the laminarian region. By splitting the rocks at fissures *Nemertes Neesii, Lineus bilineatus, Micrura fasciolata, M. purpurea*, and *Carinella annulata* are found in great beauty ; while the intricacies in the roots of the tangles afford favourite sites for *Nemertes gracilis* and others already mentioned. The débris in the fishing-boats is especially productive of fine examples of *Amphiporus pulcher* and *Micrura fusca* *, both, besides the ordinary method of progression, swimming gracefully through the water like freshwater leeches, by throwing themselves on edge and striking right and left alternately with their flattened tails. The curious *Nemertes carcinophila* is abundant on the ovigerous abdominal hairs of the females of the shore-crab.

Almost all the Nemerteans live well in confinement; and while the development of several is known, that of others (such as *Nemertes Neesii, N. gracilis, Lineus marinus, L. sanguineus*, the *Micrura*, and *Carinella annulata*) affords a fine field for further research. The Nemerteans approach the Annelids proper very closely.

The Rhabdocœla are generally minute, but tolerably numerous amongst the red ascidians hanging from cavern-roofs, or algous and zoophytic growths on the under surface of stones, in tidal pools and near low-water mark.

The Planarians are fairly represented, the common forms frequently occurring under stones between tide-marks, and

* A fine specimen of the large *Cerebratulus angulatus*, O. F. Müller, was sent me from the neighbouring Bay of Montrose by Dr. Howden : but unfortunately no proboscis was present. The two forms closely approach each other.

P

gliding over the surface of rock or glass like a living skin,
which requires a keen eye for detection. When much disturbed
they swim a short distance through the water, with a horizontal
stroke that has been compared by some to the motion of a
skate; but the undulation in the former is much greater than
in the latter, which has a gliding or skimming character.
They also progress on the surface of the water. Even more
active and irritable than the Nemerteans, they move with ease
and swiftness—never avoiding any small obstacle, but spreading their thin mobile bodies over it, and continuing their
course uninterruptedly. Occasionally when a projecting point
is attained, the anterior part of the body is elevated and waved
to and fro till a convenient branch of seaweed or zoophyte is
reached. Some are very prettily coloured; and though the
large and gaudily striped *Eurylepta vittata*, so characteristic
of our southern shores, is not found, yet the pink and yellow
hues of *Planaria ellipsis* are scarcely less attractive. The
little *Planaria ulvæ*, which abounds in the brackish waters of
many of the creeks on the western coasts, is absent. The
common *Leptoplana flexilis* may be kept for months in confinement; but it is perhaps less hardy in this respect than
the Nemerteans. Even though it perishes, however, it frequently deposits pale brownish masses of agglutinated ova on
the side of the vessel; and the development of these can easily
be followed.

Subclass TURBELLARIA.

A. **APROCTA**, Max Schultze.

Order I. DENDROCŒLA.

Fam. **Leptoplanidæ.**

Genus LEPTOPLANA, Ehrenberg.

Leptoplana subauriculata, Johnston, Catalogue of the Non-
parasitical Worms, Brit. Mus. p. 6.

Common between tide-marks.

Leptoplana flexilis, Dalyell; Johnst. Cat. p. 6.
Abundant under stones between tide-marks.

Leptoplana atomata, Müller; Johnst. Cat. p. 6.
Common in the same localities.

Leptoplana ellipsis, Dalyell; Johnst. Cat. p. 7.
Not uncommon between tide-marks.

Order II. RHABDOCŒLA.

Fam. 1. Proboscidea, J. V. Carus.

Genus PROSTOMUM, Œrst.

Prostomum lineare, Œrst.; Johnst. Cat. p. 62.
Occasionally found on stones brought from the rocks near low water.

A curious form, having a pointed snout with a globular process posteriorly, and a dull pinkish alimentary canal, was procured from the fishing-boats; but unfortunately I possess only the drawing, upon which, however, every reliance can be placed (Plate V. fig. 9).

Fam. 2. Schizostomea, O. Schm.

Genus CONVOLUTA, Œrst.

Convoluta paradoxa, Abildgaard; Johnst. Cat. p. 16.
Very common amongst seaweeds and *Corallina* in tide-pools.

Convoluta Dieslingii, Schmidt (?), Sitzungsb. der k. Akad. 1852.
Occasionally under stones in rock-pools.

Fam. 3. **Mesostomea**, O. Schm.

Genus MESOSTOMUM (Dugès), M. Sch.

Mesostomum bifidum, n. sp.

[Plate VIII. figs. 3–6.]

On the under surface of stones from low-water mark, East
Rocks. One tenth of an inch long, and of a very pretty pale
orange hue. The body is pointed anteriorly, dilates in the
middle, and diminishes posteriorly, terminating in two processes which have a few rather large papillæ on their crenated
edge; these papillæ seem to have a slight sucker-action.
There are two semilunar eyes, with the concavity external.
The cilia are specially distinct a little behind the snout, on
each side, at points corresponding to the long ciliary whips of
the developing Nemertean. The male organ formed a spirally
marked conical process behind the large median sucker; and
the testes were loaded with spermatozoa in various stages of
development.

Fam. 4. **Derostomea**, Œrst.

Genus VORTEX, Ehrenberg.

Vortex capitata, Œrst. Entwurf Plattwürmer, p. 65,
pl. 1. f. 7.

[Plate VIII. figs. 7–10.]

Occasionally under stones between tide-marks. Many dia-
toms occur in the digestive canal.

Fam. 5. **Opistomea**, O. Schm.

Genus MONOCELIS, Ehrb.

Monocelis unipunctata (Fab.), Œrst. Ent. Plattw. p. 56.

This appears to be the *Planaria flustræ* of Dalyell. It is
abundant under stones between tide-marks.

Monocelis rutilans, O. F. Müller, Zool. Danic. iii. p. 49,
tab. 109. f. 10 & 11.

Occasionally in the laminarian region.

Order NEMERTINEA.

Suborder **ENOPLA.**
Proboscis furnished with stylets.

Fam. 1. **Amphiporidæ.**

Subfamily *AMPHIPORINÆ.*
Proboscis proportionally large.

Genus 1. AMPHIPORUS, Ehrenberg.

Amphiporus lactifloreus, Johnst. M'Intosh, Brit. Annel.
    (Ray Society), i. p. 156, pl. 1. f. 1 & 2.

Common under stones between tide-marks.

Amphiporus pulcher (O. F. Müller), Johnst. *Op. cit.* p. 158,
    pl. 1. f. 3.

Frequent in the coralline ground in crevices of shells. A
very large, though fragmentary, specimen appears to be this
species (Plate IV. fig. 3). It was found by its artist on
the West Sands after a severe storm in March. The proboscis is extruded, and the tip of the snout forms a kind
of button, which, however, may be due to the condition of
the parts ; the œsophageal region protrudes as a rugose disk
on the ventral surface near the tip. A reddish line along
the body is peculiar, and is probably the nerve-cord, since
no vivid colouring has been seen in the vessels of ordinary
specimens. The dull greyish coloration of the body is also
peculiar, and may be partly owing to the brownish-red ova
shining through the other tissues, or to alimentary material.
The posterior end of the specimen shows the pinkish elements
of the digestive chamber and ova.

Genus 2. TETRASTEMMA, Ehrenberg.

Tetrastemma melanocephala, Johnst. *Op. cit.* p. 165, pl. 2. f. 1.

Not rare amongst the roots of seaweeds on stones near low-
water mark.

Tetrastemma candida, O. F. Müller.   *Op. cit.* p. 167,
pl. 2. f. 2 & 3.

Abundant amongst seaweeds on stones in the same localities.

Tetrastemma vermicula, De Quatrefages.   *Op. cit.* p. 169,
pl. 3. f. 3.

Common amongst the roots of seaweeds on stones between
tide-marks.

Tetrastemma flavida, Ehrenberg.   *Op. cit.* p. 170, pl. 4. f. 1.

Not uncommon in the same situations.

Tetrastemma dorsalis, Abildgaard.   *Op. cit.* p. 172, pl. 1. f. 4,
& pl. 3. f. 4.

In swarms on *Ceramium* and other seaweeds in the laminarian region, and occasionally under stones near low-water
mark.

Subfamily *NEMERTINÆ*.
Proboscis proportionally small.

Genus 4. NEMERTES, Cuvier.

Nemertes gracilis, Johnst.   *Op. cit.* p. 176, pl. 2. f. 5.

Frequent under tangle-roots at low water, and occasionally
under stones between tide-marks.

Nemertes Neesii, Œrst.   *Op. cit.* p. 178, pl. 3. f. 6,
& pl. 7. f. 6.

Common in the same localities, in fissures of the rocks between tide-marks, and often from deep water.

Nemertes carcinophila, Kölliker.   *Op. cit.* p. 180, pl. 1. f. 5.

Very frequent on the abdominal hairs of female *Carcini*.

Genus 5. LINEUS, Sowerby.

Lineus marinus, Montagu.   *Op. cit.* p. 181, pl. 9.

Common between tide-marks and in deep water.

Lineus gesserensis, O. F. Müller.   *Op. cit.* p. 185, pl. 4. f. 2,
           & pl. 5. f. 1.
Abundant between tide-marks.   Green and red varieties
are equally common.

Lineus sanguineus, Jens Rathke.   *Op. cit.* p. 188, pl. 5. f. 2.
Somewhat less common than the former, in the same sites.

Lineus bilineatus, Delle Chiaje.   *Op. cit.* p. 191, pl. 6. f. 1.
Not uncommon between tide-marks, and in deep water.

Genus 8. MICRURA, Ehrenberg.
Micrura fusca, M'Intosh.   *Op. cit.* p. 196, pl. 6. f. 3.
Common in the coralline ground amongst old shells.

Micrura fasciolata, Ehrenberg.   *Op. cit.* p. 197, pl. 6. f. 2.
Not rare in fissures of the rocks between tide-marks, and
occasionally from deep water.   The uniformly tinted variety
frequents the latter.

Micrura purpurea, Dalyell.   *Op. cit.* p. 200, pl. 7. f. 3.
Occasionally in the same localities.

Fam. 3. Carinellidæ.

Genus 10. CARINELLA, Johnst.
Carinella annulata, Montagu.   *Op. cit.* p. 203, pl. 7. f. 5,
           & pl. 8.
Common between tide-marks, and in deep water amongst
shells.

Fam. 4. Cephalotrichidæ.

Genus 12. CEPHALOTHRIX, Œrst.
Cephalothrix linearis, Jens Rathke.   *Op. cit.* p. 208,
           pl. 4. f. 4 & 5.
Abundant under muddy stones between tide-marks.

Subclass CHÆTOGNATHA.

Genus SAGITTA, Slabb.

Sagitta bipunctata, Quoy & Gaimard (?), Krohn.

Vast numbers were found on the West Sands, after a severe storm, in January 1867. They were scattered amidst the foam on the beach along with multitudes of *Pleurobrachia*; and it is curious that very little else was cast ashore at this time. The season is remarkable, as Prof. Busk, who is the author of a most valuable paper * on the structure and relations of the animal, thought it would chiefly be procured in fine and calm weather in the towing-net. They were recognized by the active movements of their bodies, which glistened all along the beach like needles of glass.

Subclass GEPHYREA.

Fam. Echiuridea, J. V. Carus.

Genus ECHIURUS, Cuvier.

Echiurus vulgaris, Sav.; Baird, Proc. Zool. Soc. 1868, p. 109.

[Plate IV. fig. 1.]

Abundant amongst the débris on the West Sands after storms.

Fam. Sipunculidea, J. V. Carus.

Genus PHASCOLOSOMA, F. S. Leuck.

Phascolosoma Harveii, Forbes; Baird, *loc. cit.* p. 82.

Abundant in the stomachs of cod and haddock.

Phascolosoma Strombi, Montagu; Baird, *loc. cit.* p. 86.

Common in deep water in *Dentalium, Turritella,* and *Aporrhais.*

* Journ. of Microscop. Science, 1856, p. 14.

Phascolosoma Johnstoni, Forbes ; Baird, *loc. cit.* p. 95.

Frequent amongst the roots of corallines and seaweeds on stones in pools, and in crevices of rocks. Ranges to deep water in shells.

Fam. **Priapulidea**, J. V. Carus.

Genus PRIAPULUS, Lam.

Priapulus caudatus, Lam. ; Baird, *loc. cit.* p. 104.

[Plate IV. fig. 2.]

Plentiful in the stomachs of cod and haddock, and from deep water.

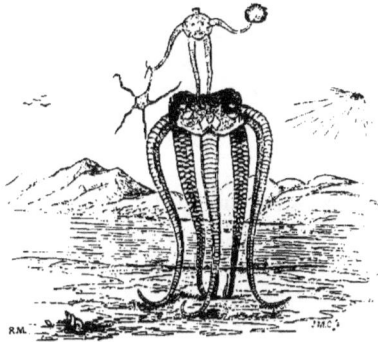

R.M.   J.M.C.

q

Subclass ANNULATA DISCOPHORA.

Fam. 1. Hirudinea, Savigny.

Genus PONTOBDELLA, Leach.

Pontobdella muricata, L.; Johnst. Cat. p. 39.

[Plate V. fig. 1.]

Abundant on skate, and tossed on the West Sands after storms.

Pontobdella littoralis, Johnst. Cat. p. 42.

[Plate V. figs. 3–6.]

Not uncommon on *Cottus bubalis* thrown on the West Sands after storms, and occasionally in the stomach of the haddock. It is curious that an example of *Piscicola geometra* (Plate V. fig. 2) should have been found on the former fish on the sands near the mouth of the Eden.

Fam. 5. Malacobdellea, J. V. Carus.

Genus MALACOBDELLA, Blainville.

Malacobdella grossa, O. F. Müller; Johnst. Cat. p. 35.

Occasionally in *Cyprina islandica*. The late Dr. Fraser Thomson procured my specimen.

Subclass ANNULATA OLIGOCHÆTA.

Fam. Lumbricina, D'Ud.

Genus CLITELLIO, Sav.

Clitellio arenarius, O. F. Müller; Johnst. Cat. p. 66.

In swarms under stones on sandy and muddy ground between tide-marks.

Subclass ANNULATA POLYCHÆTA.

Fam. 3. Aphroditidæ.

Genus APHRODITA, L.

Aphrodita aculeata, L.; Malmgren, Annulata Polychæta &c. p. 3.

Abundant in deep water, and thrown on the West Sands in thousands after some winter storms. It is also a common diet of the cod and haddock.

Fam. 4. Polynoidæ.

Genus LEPIDONOTUS, Leach.

Lepidonotus squamatus, L.; Mgrn. *op. cit.* p. 4.

Frequent in deep water, under stones in pools between tide-marks, on the West Sands after storms, and in the stomachs of cod and haddock.

Genus NYCHIA, Mgrn.

Nychia cirrosa, Pallas; Mgrn. *op. cit.* p. 5.

Occasionally in deep water, and on the West Sands after storms.

Genus LAGISCA, Mgrn.

Lagisca propinqua, Mgrn. *op. cit.* p. 9.

Occasionally in débris of the fishing-boats. It is distinguished by its greyish scales mottled with black, by the dark spots at the bases of the feet, the mottling of the dorsum beneath the scales, and by the position of the eyes (the posterior pair only being visible from the dorsum). The dorsal bristles have a short clear portion at the tip; the ventral are long, much tapered and minutely bifid superiorly, while the inferior have shorter and stouter tips, more evidently bifid.

Genus HARMOTHOË, Kinberg.

Harmothoë imbricata, L.; Mgrn. *op. cit.* p. 9.

Very abundant under stones between tide-marks, and ranging to deep water.

Harmothoë lunulata, Delle Chiaje, Descriz. e Not. pl. 144. f. 5 & 6 (*fide* Claparède).

Occasionally on the West Sands after storms.

Harmothoë Macleodi, M'Intosh.

Stomach of the cod. It is allied to *H. zetlandica* in regard to general appearance and processes. Scales fourteen to fifteen pairs, pale and semitranslucent; dorsal cirri scarcely extend beyond the bristles; serrations of the dorsal bristles continued to the tip; ventral bristles boldly bifid, and with rather broad tips.

Genus POLYNOË, Sav.

Polynoë floccosa, Savigny, Syst. des Annél. p. 23.

Not uncommon on the West Sands after storms, and under stones between tide-marks.

Genus EVARNE, Mgrn.

Evarne impar, Johnst.; Mgrn. *op. cit.* p. 10.

Occasionally under stones in pools between tide-marks, and in littoral sponges.

Genus LÆNILLA, Mgrn.

Lænilla setosissima, Savigny, Syst. des Annél. p. 25; Mgrn. *op. cit.* p. 12.

Polynoë longisetis, Grube; *Lænilla glabra*, Mgrn.; and *Harmothoë Malmgreni*, Lankester.

Tossed on the West Sands after storms, amongst tangle-roots. Not uncommon.

Genus HERMADION, Kinberg.

Hermadion pellucidum, Ehlers, Die Borstenwürmer, i. p. 105, pls. 3 & 4.

Occasionally in deep water amongst corallines and shells.

Hermadion assimile, M'Intosh.

Amongst the débris in the fishing-boats. This species is easily discriminated from the foregoing (in spirit) by the presence of a brownish-black band commencing behind the head, and continuing along the central line to the tail. Dorsal bristles with the rows of spikes much less marked, and with a notch at the tip of each bristle; the ventral bristles have a somewhat blunt tip, with processes or beaks which differ characteristically from those of the foregoing.

Genus HALOSYDNA, Kinberg.

Halosydna gelatinosa, Sars; Mgrn. *op. cit.* p. 14.

Not uncommon under stones in rock-pools and in the stomach of the cod.

Genus MALMGRENIA, M'Intosh.

Malmgrenia andreapolis, M'Intosh.

Amongst the débris in the fishing-boats, in the stomachs of cod and haddock, and abundantly on the West Sands after storms. The scales have a persistent brown belt. Dorsal bristles terminated by a peculiar knob; ventral bifid, but the distal process is constituted by a modification of the knob.

Genus ENIPO, Mgrn.

Enipo Kinbergi, Mgrn. *op. cit.* p. 15.

Occasionally in the stomachs of cod and haddock.

Fam. 6. Sigalionidæ.

Genus STHENELAIS, Kinberg.

Sthenelais boa, Johnst. Cat. Brit. Mus. p. 124.

Not uncommon between tide-marks under stones.

Sthenelais limicola, Ehlers, Die Borstenwürmer, i. p. 120,
pls. 4 & 5.

Abundant on the West Sands after storms, and in the stomachs of cod, haddock, and flounders.

Genus SIGALION, M.-Edwards.

Sigalion Mathildæ, M.-Ed. Hist. du Litt. de la France, ii.
p. 105, pl. 2.

Common on the West Sands after storms, and in the stomachs of cod and haddock.

Genus PHOLOË, Johnst.

Pholoë minuta, Fab.; Mgrn. *op. cit.* p. 17.

Frequent under stones between tide-marks, and also in deep water.

Fam. 7. Nephthydidæ.

Genus NEPHTHYS, Cuvier.

Nephthys cœca, Fab.; Mgrn. *op. cit.* p. 18.

Common on the beach after storms, in sand under stones between tide-marks, and in the stomachs of cod, haddock, and other fishes.

Nephthys Hombergii, And. & M.-Ed. Hist. Litt. *olim cit.*
p. 235, pl. 5 b. f. 1–6.

Not uncommon between tide-marks, and in the stomachs of cod and haddock.

Nephthys Johnstoni (*longisetosa*, Johnst.).

Occasionally between tide-marks, and in the stomachs of cod and haddock. This quite differs from the *N. longisetosa* of Œrsted, Malmgren, and others.

Fam. 8. Phyllodocidæ.

Genus NOTOPHYLLUM, Œrst.

Notophyllum foliosum, Sars; Mgrn. *op. cit.* p. 19.

Amongst the débris of the fishing-boats. Not common.

Genus GENETYLLIS, Mgrn.

Genetyllis lutea, Mgrn. *op. cit.* p. 20.

Occasionally in deep water.

Genus PHYLLODOCE, Savigny.

Phyllodoce grœnlandica, Œrst.; Mgrn. *op. cit.* p. 21.

Thrown in numbers on the West Sands after storms; stomachs of cod and haddock.

Phyllodoce maculata, O. F. Müller ; Johnst. Cat. p. 177.

Common under stones between tide-marks.

Phyllodoce laminosa, Savigny ; Mgrn. *op. cit.* p. 24.

Frequent between tide-marks, in the laminarian region, on the West Sands after storms, and in the stomachs of cod, haddock, and other fishes.

Genus EUMIDA, Mgrn.

Eumida sanguinea, Œrst., and var.; Mgrn. *op. cit.* p. 25.

Common between tide-marks under stones, and on the West Sands after storms.

Genus EULALIA (Sav.), Mgrn.

Eulalia viridis, O. F. Müller; Mgrn. *op. cit.* p. 25.

Abundant between tide-marks, and ranging to deep water.

Eulalia bilineata, Johnst.; Mgrn. *op. cit.* p. 25.

Under stones in rock-pools. Frequent.

Eulalia tripunctata, n. sp.

Amongst the débris of the fishing-boats. Colour pale yellow, with three rows of black spots on the dorsum.

Genus ETEONE (Sav.), Mgrn.

Eteone picta, De Quatref. Annelés, ii. p. 147.

West Sands after storms, in the stomachs of cod and haddock, and rarely under stones at the East Rocks.

Eteone andreapolis, M'Intosh.

On the West Sands after storms. The species has large eyes, a peculiarly shaped head, and madder-brown or purplish bands on the dorsum.

Eteone arctica, Mgrn. (?); Mgrn. *op. cit.* p. 27.

West Sands after storms.

Genus ETEONELLA, M'Intosh.

Eteonella Robertianæ, M'Intosh.

Found whilst digging for littoral annelids. It appears to be most closely allied to *Eteone longa*, Œrsted. Head conical, with a distinct furrow on each side; and, like Malmgren's *Chætoparia*, the cephalic and buccal segments seem to be united; for two short filiform tentacles proceed from the posterior part of the head; the mouth, moreover, opens in the cephalic segment; no visible eyes in spirit; there is a distinct elevation in the centre of the head posteriorly; the lobes of the feet are lanceolate.

Fam. 9. Hesionidæ.

Genus CASTALIA, Savigny.

Castalia punctata, O. F. Müller; Mgrn. *op. cit.* p. 31.

Not uncommon in deep water, and occasionally under stones near low-water mark.

Genus PSAMATHE, Johnst.

Psamathe fusca, Johnst. Cat. Brit. Mus. p. 182, pl. 14 a. f. 4.

Frequent under stones in pools and moist places between tide-marks.

Fam. 10. Syllidæ.

Genus AUTOLYTUS, Grube.

Autolytus prolifer, O. F. Müller; Mgrn. *op. cit.* p. 32.

Not uncommon near low-water mark under stones, and ranging to deep water.

Autolytus (Proceræa) pictus, Ehlers, Die Borstenwürmer, i. p. 256, pl. 11. f. 8–17.

Occasionally under stones in tide-pools. This form also appears to show alternation of generations.

Genus EUSYLLIS, Mgrn.

Eusyllis tubifex, Gosse (?); M'Intosh, Trans. Roy. Soc. Edinb. vol. xxv. 2. p. 414.

Abundant on laminarian blades cast ashore by storms.

Genus EXOGONE, Œrst. (*Sphærosyllis*).

Exogone naidina, Œrst. (?), Archiv f. Naturg. xi. 1845, p. 20, Taf. 2.

Occasionally under stones in rock-pools.

Genus SYLLIS, Savigny.

Syllis armillaris, O. F. Müller; Mgrn. *op. cit.* p. 42.

Frequent between tide-marks under stones, and in the laminarian region.

R

Genus " IOIDA," Johnst.

"*Ioida macrophthalma*," Johnst. Cat. Brit. Mus. p. 197,
pl. 14 *a*. f. 5.

Occasionally between tide-marks. This is the sexual bud
of a *Syllis*. Four or five of the segments anteriorly are devoid
of the long bristles.

Fam. 11. Nereidæ.

Genus NEREIS, L.

Nereis pelagica, L.; Mgrn. *op. cit.* p. 47.

Everywhere abundant from high-water mark to the coralline
ground, and in the stomachs of many fishes.

Nereis cultrifera, Grube; Ehlers, Die Borstenwürmer,
ii. p. 461, pls. 18–20.

Frequent between tide-marks under stones on muddy
ground, and in the stomachs of various fishes.

Nereis Dumerilii, Aud. & M.-Ed.; Ehlers, *op. cit.* p. 535.

It is curious that only the epitocous form (olim *Iphinereis
fucicola*) has yet occurred, viz. in the coralline region and on
the West Sands after storms.

Genus HEDISTE, Mgrn.

Hediste diversicolor, O. F. Müller; Mgrn. *op. cit.* p. 49.

Occasionally between tide-marks, and after storms on the
West Sands.

Genus EUNEREIS, Mgrn.

Eunereis longissima, Johnst.; Mgrn. *op. cit.* p. 57.

Occasionally cast ashore on the West Sands after storms.
This is an epitocous form, the relations of which are at present
in obscurity; there is no known species with which it may be
connected except those mentioned here.

Genus NEREILÆPAS, Blainville.

Nereilepas fucata, Savigny; Mgrn. *op. cit.* p. 53.

Abundant on the coralline ground, chiefly in company with *Pagurus* in *Buccinum.* It also occurs in the stomachs of various fishes.

Genus ALITTA, Kinberg.

Alitta virens, Sars; Mgrn. *op. cit.* p. 56.

Sometimes thrown in large numbers on the West Sands after storms, and not uncommon in the stomachs of cod.

Fam. 13. **Lumbriconereidæ.**

Genus LUMBRICONEREIS (Blainv.), M.-Edwards.

Lumbriconereis fragilis, O. F. Müller; Mgrn. *op. cit.* p. 63.

West Sands after storms, and in the stomachs of haddock and flounders. Not rare.

Lumbriconereis Laurentiana, Grube, Archiv f. Naturg.
Bd. xxix. 1863, p. 40.

Stomachs of cod and haddock.

Fam. 15. **Onuphididæ.**

Genus ONUPHIS, Sars.

Onuphis tubicola, O. F. Müller; Mgrn. *op. cit.* p. 67.

Fragmentary specimen in the stomach of a haddock.

Fam. 16. **Goniadidæ.**

Genus GONIADA, And. & M.-Ed.

Goniada maculata, Œrst.; Mgrn. *op. cit.* p. 68.

Common in the stomachs of cod and haddock.

R 2

Fam. 17. Glyceridæ.

Genus GLYCERA, Savigny.

Glycera dubia, Blainv. (vel *Rouxii*, Aud. & M.-Ed. ?).
West Sands after storms and in fissures of rocks.

Glycera capitata, Œrst. ; Mgrn. *op. cit.* p. 70.
Occasionally in the stomachs of cod and haddock.

Glycera Goësi, Mgrn. *op. cit.* p. 71.
Stomachs of cod, haddock, and flounders.   Not uncommon.

Fam. 18. Ariciidæ.

Genus ARICIA, Savigny.

Aricia Cuvieri, Aud. & M.-Ed. ; Mgrn. *op. cit.* p. 71.
Common between tide-marks in sand, and thrown on the
West Sands after storms.

Genus SCOLOPLOS (Blainv.), Œrst.

Scoloplos armiger, O. F. Müller ; Mgrn. *op. cit.* p. 72.
Frequent between tide-marks under stones on sandy ground.

Fam. 19. Opheliidæ.

Genus AMMOTRYPANE, H. Rathke.

Ammotrypane aulogaster, H. Rathke ; Mgrn. *op. cit.* p. 73.
Occasionally in the stomachs of haddocks.

Genus OPHELIA (Sav.), M.-Edwards.

Ophelia limacina, H. Rathke ; Mgrn. *op. cit.* p. 74.
Very abundant on the West Sands after storms, and often
in the stomachs of cod and haddock.

Genus TRAVISIA, Johnst.

Travisia Forbesii, Johnst.; Mgrn. *op. cit.* p. 75.
Occasionally under stones on gravel at East Rocks, and in
the stomachs of flounders.

Fam. 20. Scalibregmidæ.

Genus EUMENIA, Œrst.

Eumenia crassa, Œrst.; Mgrn. *op. cit.* p. 76.
In the stomach of the haddock. Not uncommon.

Genus SCALIBREGMA, H. Rathke.

Scalibregma inflata, H. Rathke; Mgrn. *op. cit.* p. 77.
In the stomach of a flounder. Rare.

Fam. 21. Telethusidæ.

Genus ARENICOLA, Lamarck.

Arenicola marina, L.; Mgrn. *op. cit.* p. 78.
Everywhere abundant in sandy ground.

Fam. 22. Sphærodoridæ.

Genus EPHESIA, H. Rathke.

Ephesia gracilis, H. Rathke ; Mgrn. *op. cit.* p. 79.
Occasionally between tide-marks, and frequently in the
coralline region.

Fam. 23. Chloræmidæ.

Genus TROPHONIA, M.-Edwards.

Trophonia plumosa, O. F. Müller; Mgrn. *op. cit.* p. 82.
Common on the beach after storms, in muddy fissures of
the rocks between tide-marks, and ranging to deep water, as
well as in the stomachs of various fishes.

Genus FLABELLIGERA, Sars.

Flabelligera affinis, Sars; Mgrn. *op. cit.* p. 83.

Frequent in deep water, and sometimes between tide-marks.

Fam. 25. Chætopteridæ.

Genus CHÆTOPTERUS, Cuvier.

Chætopterus norvegicus, Sars; Mgrn. *op. cit.* p. 88.

Occasionally in the stomachs of haddock.

Genus MÆA, Johnst.

Mæa mirabilis, Johnst. Cat. Brit. Mus. p. 278.

Not uncommon amongst gravelly sand off the East Rocks. The position of this remarkable form may be regarded as provisional (between the Chætopteridæ and Spionidæ).

Fam. 26. Spionidæ.

Genus NERINE, Johnst.

Nerine foliosa, Sars; Mgrn. *op. cit.* p. 89.

Common in sandy ground or in muddy sand.

Genus SCOLECOLEPIS, Blainv.

Scolecolepis vulgaris, Johnst.; Mgrn. *op. cit.* p. 90.

Not uncommon in muddy sand.

Genus SPIO, Œrst.

Spio seticornis, Fab.; Mgrn. *op. cit.* p. 92.

In fine sand tubes under stones at the East Rocks.

Genus POLYDORA, Bosc.

Polydora ciliata, Johnst. ; Mgrn. *op. cit.* p. 95.

Very abundant in soft sandstone and shale.

Fam. 27. Cirratulidæ.

Genus CIRRATULUS, Lamarck.

Cirratulus cirratus, O. F. Müller; Mgrn. *op. cit.* p. 95.

Common in mud and muddy sand under stones between tide-marks, and ranging to deep water.

Genus DODECACERIA, Œrst.

Dodecaceria concharum, Œrst.; Mgrn. *op. cit.* p. 96.

Not unfrequent in tangle-roots and old shells from low-water mark to the coralline ground.

Fam. 28. Capitellidæ.

Genus CAPITELLA, Blainville.

Capitella capitata, Fab.; Mgrn. *op. cit.* p. 97.

Common on the West Sands after storms and in fissures of rocks in mud.

Fam. 29. Maldanidæ.

Genus MALDANE, Grube.

Maldane biceps, Sars; Mgrn. *op. cit.* p. 98.

A fragmentary specimen in the stomach of a haddock.

Genus NICOMACHE, Mgrn.

Nicomache lumbricalis, Fabr.; Mgrn. *op. cit.* p. 99.

Common between tide-marks under stones, and thence to deep water; it is especially abundant in vertical fissures of the soft sandstone at the East Rocks.

Genus PRAXILLA, Mgrn.

Praxilla prætermissa, Mgrn. *op. cit.* p. 100.
In the stomach of a haddock.   Not common.

Fam. 30. **Ammocharidæ.**

Genus OWENIA, Delle Chiaje.

Owenia filiformis, Delle Chiaje ; Claparède, Chæt. Naples,
p. 446, pl. 26. f. 5.
Common in the stomachs of haddock.

Fam. 31. **Hermellidæ.**

Genus SABELLARIA, Lamarck.

Sabellaria spinulosa, R. Leuckart ; Mgrn. *op. cit.* p. 102.
Abundant between tide-marks, and thence to deep water.

Fam. 32. **Amphictenidæ.**

Genus PECTINARIA, Lamarck.

Pectinaria belgica, Pallas ; Mgrn. *op. cit.* p. 103.
Very abundant off the West Sands, and tossed ashore in
vast numbers after storms.   Common in the stomachs of cod
and haddock.

Genus AMPHICTENE, Sav.

Amphictene auricoma, O. F. Müller ; Mgrn. *op. cit.* p. 103.
Occasionally off the East Rocks in sandy ground, and in
the stomachs of cod, haddock, and flounders.

Fam. 33. **Ampharetidæ.**

Genus AMPHARETE, Mgrn.

Ampharete arctica, Mgrn. *op. cit.* p. 104.
Occasionally in deep water and in the stomachs of haddock.

Genus AMPHICTEIS (Gr.), Mgrn.

Amphicteis Gunneri, Sars; Mgrn. *op. cit.* p. 105.

Not uncommon in the stomachs of haddock.

Genus MELINNA, Mgrn.

Melinna cristata, Sars; Mgrn. *op. cit.* p. 106.

Frequent in the stomachs of cod.

Fam. 34. Terebellidæ.

Subfam. 1. *AMPHITRITEA*, Mgrn.

Genus AMPHITRITE, O. F. Müller.

Amphitrite figulus, Dalyell; Mgrn. *op. cit.* p. 107 (as *A. Johnstoni*).

Not uncommon between tide-marks, and ranging to deep water.

Genus LANICE, Mgrn.

Lanice conchilega, Pallas; Mgrn. *op. cit.* p. 108.

Abundant between tide-marks and off the West Sands, and multitudes are thrown on the beach after storms. A common food of many fishes.

Genus NICOLEA, Mgrn.

Nicolea zostericola, Œrst. & Gr.; Mgrn. *op. cit.* p. 109.

Common between tide-marks amongst tangle-roots, and ranging to deep water.

Genus THELEPUS, Leuckart.

Thelepus circinatus, Fab.; Mgrn. *op. cit.* p. 110.

Frequent in the laminarian and coralline regions, in the stomachs of various fishes, and on the West Sands after storms.

s

Subfam. 2. *POLYCIRRIDEA*, Mgrn.

Genus POLYCIRRUS, Grube.

Polycirrus (Ereutho) Smitti, Mgrn. *op. cit.* p. 111.
Not uncommon between tide-marks.

Subfam. 5. *CANEPHORIDEA*, Mgrn.

Genus TEREBELLIDES, Sars.

Terebellides Strœmii, Sars ; Mgrn. *op. cit.* p. 112.
Large specimens occur in the stomachs of cod and haddock.

Fam. 35. Sabellidæ.

Genus SABELLA, L.

Sabella pavonia, Sav. ; Mgrn. *op. cit.* p. 112.
Abundant in the coralline ground, on the West Sands after
storms, and in the stomach of the cod.

Sabella (Branchionima, Kölliker) *vesiculosa*, Mont. ; Johnst.
Cat. Brit. Mus. p. 259.
Frequently thrown on the West Sands after storms.

Sabella viridis, M.-Edwards, Règ. An. Illust. pl. 1 e
(*fide* De Quatref.).
Amongst mud in the interstices of *Filigrana implexa* from
the coralline region.

Genus DASYCHONE, Sars.

Dasychone Dalyelli, Kölliker ; Mgrn. *op. cit.* p. 115.
Occasionally from the coralline ground in the débris of
fishing-boats.

Genus Amphicora, Ehrenberg.

Amphicora Fabricia, O. F. Müller; Mgrn. *op. cit.* p. 117.

Abundant under stones on muddy ground between tide-marks and amongst tangle-roots.

.

Fam. 36. **Serpulidæ.**

Genus Protula, Risso.

Protula tubularia, Mont. (=*protensa*, Johnst.); Johnst. Cat. Brit. Mus. p. 264.

Occasionally in deep water.

Genus Filigrana, Oken.

Filigrana implexa, Berkeley; Mgrn. *op. cit.* p. 119.

Fine masses are common in the coralline region.

Genus Hydroides, Gunner.

Hydroides norvegica, Gunner; Mgrn. *op. cit.* p. 120.

Abundant in deep water, attached to shells, stones, &c.

Genus Serpula, L.

Serpula vermicularis, L.; Mgrn. *op. cit.* p. 120.

Common in deep water.

Genus Pomatocerus, Phil.

Pomatocerus triqueter, L.; Mgrn. *op. cit.* p. 121.

Very common from the littoral to the coralline region.

Genus SPIRORBIS, Daud.

Spirorbis borealis, Daud. ; Mgrn. *op. cit.* p. 122.

Abundant on seaweeds and stones between tide-marks.

Spirorbis lucidus, Mont. ; Mgrn. *op. cit.* p. 123.

Common on zoophytes from deep water.

Series II. ARTHROPODA.

Class CRUSTACEA.

The sessile-eyed Crustacea of St. Andrews are tolerably numerous both in species and individuals. Between tide-marks the most conspicuous (as usual) are the swarms of *Talitrus locusta* which speedily reduce dead fish and other animals to skeletons at high-water mark and considerably beyond it, and the multitudes of *Gammarus locusta* and *Amphithoë podoceroides* under stones amongst the rocks. The *Podocerides*, *Pherusa bicuspis*, *Calliopius grandoculis*, and *Caprella tuberculata* are plentiful in the rock-pools, and *Corophium grossipes* in the brackish pools near the estuary of the Eden. *Janira maculosa* abounds both in the tidal region and in deep water, while *Jæra Nordmanni* occurs in numbers under stones near high-water mark. In the laminarian region one of the most abundant, perhaps, is *Atylus Swammerdami*, which congregates in swarms on the loose seaweeds. *Siphonœcetus typicus* is common amongst shell-gravel, and *Eurydice pulchra* on the surface of the sea as well as in rock-pools in autumn. Many of the rarer forms occur in the deeper water in considerable numbers; but the distribution of the group in British seas is still involved in considerable obscurity; and at present it will suffice to observe that two of the most plentiful in this region are *Ampelisca Belliana*, Bate, and the new *Calliopius bidentatus*, Norman. The former is likewise common on the beach after storms and in the stomachs of fishes; and the latter ranges to the laminarian zone.

Compared with the Zetlandic area, the absence at St. Andrews of such forms as *Acanthonotus Owenii*, *Dexamine cedlomensis*, *Cymodocea truncata*, and *Sphæroma Prideauxianum* in the laminarian region strikes even a superficial observer of the group; while the large number of rare and new species which were met with during the frequent dredgings of Dr. Gwyn Jeffreys and the Rev. A. M. Norman still further heightens the contrast. The southern region, again, is boldly separated by the presence in considerable numbers of *Cymodocea truncata* and *Sphæroma Prideauxianum* in the fissures of rocks between tide-marks, and *Dynamene* in rock-pools. The characteristic *Tanais vittatus, Paranthura costana, Næsa bidentata, Mœra grossimana, Chelura terebrans, Conilera cylindrica,* and the large *Cymothoa* parasitic on the fishes at once distinguish the fauna of the Channel Islands from that at St. Andrews. The rarity of *Orchestia littorea* at the latter and its abundance in the tidal region of the Outer Hebrides, and the absence of *Sulcator arenarius* and its frequent occurrence in the sand of the western shores of England, are also interesting contrasts.

Many of the sessile-eyed Crustacea, such as *Talitrus locusta,* are extremely hardy. *Gammarus locusta* is often found in putrid localities, and it survives almost every other marine form in putrid vessels in confinement. The group as a whole is composed of extremely active animals; and even the most grotesque, such as *Caprella tuberculata,* are at home in the intricacies of *Ceramium* and other finely branched seaweeds. The boring forms (by jaws) are represented by *Limnoria lignorum*; but its depredations are comparatively insignificant, probably because little wood is employed within water-mark in the construction of the harbour. The perforations of *Talitrus,* again, abound in the sand, and the looped burrows of *Corophium* in the sandy mud of the flats it inhabits. The nest-forming crustaceans are represented by *Amphithoï podoceroides, Siphonœcetus typicus, Podocerus variegatus,* and *P. falcatus*; while the young of *Gammarus locusta* are often observed adhering to the abdominal region of the parent.

The Cirripedes occur abundantly between tide-marks, the most conspicuous being *Balanus balanoides,* which covers the bare rocky ridges opposite the Castle and other parts. In deep water the various species are attached to shells, stones, crabs, wood, cork, coal, tests of ascidians, and other structures.

I am indebted to Mr. Spence Bate for the determination of several doubtful forms, and especially to the Rev. A. M. Norman for his courteous assistance in this respect, and in revising the list. Mr. G. S. Brady kindly furnished me with the names of the Ostracoda occurring in shell-débris on the West Sands and other collections.

Order Pycnogonoidea.

Fam. Pycnogonidæ, Latreille.

Genus Pycnogonum, Brünnich.

Pycnogonum littorale, O. F. Müller.

Abundant under stones between tide-marks.

Genus Phoxichilidium, M.-Edwards.

Phoxichilidium femoratum, Rathke.

Occasionally under stones in rock-pools, and ranging to deep water.

Besides the foregoing, there are several species (one apparently identical with Mr. Goodsir's *Nymphon Johnstoni,* and another with his *N. spinosum*) not uncommon in the coralline region. Many delicate zoophytes are found on their limbs.

Order Cirripedia.

Suborder SUCTORIA.

Fam. Peltogastridæ, Claus.

Genus Peltogaster, H. Rathke.

Peltogaster paguri, H. Rathke.

Occasionally on *Pagurus bernhardus.* A more elongated form occurs on *P. cuanensis.*

Genus Sacculina, Thompson.

Sacculina carcini, Thompson.

[Plate IX. fig. 13.]

Common on the abdomen of *Carcinus mœnas.* Another is found on *Portunus holsatus* (Plate IX. figs. 14 & 15).

Suborder **THORACICA**.

Fam. **Lepadidæ**.

Genus LEPAS, L.

Lepas anatifera, L. ; Darwin, Mon. i. p. 73, pl. 1. f. 1.

On the bottoms of ships, and thrown ashore after storms attached to timber.

Genus SCALPELLUM, Leach.

Scalpellum vulgare, Leach ; Darw. Mon. i. p. 222, pl. 5. f. 15.

On *Thuiaria thuja* and *Sertularia cupressina* from deep water.

Fam. **Balanidæ**.

Subfamily *BALANINÆ*.

Genus BALANUS, Lister.

Balanus porcatus, E. da Costa ; Darw. Mon. ii. p. 256, pl. 6. f. 4.

Abundant on stones, *Ascidia sordida*, crabs, &c. in deep water, and occasionally between tide-marks.

Balanus crenatus, Bruguière; Darw. Mon. ii. p. 261, pl. 6. f. 6.

Not uncommon on *Hyas araneus*, *Lithodes maia*, and on rocks in the laminarian region.

Balanus balanoïdes, L. ; Darw. Mon. ii. p. 267, pl. 7. f. 2.

Very abundant ; coating extensive surfaces of the rocks between tide-marks and in the laminarian region, and adhering to mussels, sticks, posts, &c. Elongated varieties are not uncommon. The exuviæ swarm in the rock-pools and on the surface of the sea in summer.

Balanus Hameri, Ascanius ; Darw. Mon. ii. p. 277, pl. 7. f. 5.

Occasionally in deep water ; a small thorn-tree (still fresh) was covered with fine examples.

Fam. Verrucidæ.

Genus VERRUCA, Schumacher.

Verruca Strömii, O. F. Müller; Darw. Mon. p. 518, pl. 21. f. 1.

Abundant on rocks and stones between tide-marks in the laminarian region, and on crabs in the coralline.

Order COPEPODA.

Suborder GNATHOSTOMATA.

Genus NOTODELPHYS, Allman.

Notodelphys ascidicola, Allman.

Common in *Ascidia intestinalis* and others.

Suborder PARASITA.

Genus CALIGUS, O. F. Müller.

Caligus rapax, M.-Edwards.

Common on cod. Many specimens have *Udonella caligorum* attached to them. Free specimens often occur in rock-pools.

Genus LEPEOPHTHEIRUS, Nordm.

Lepeophtheirus salmonis, Kröyer.

Abundant on the salmon.

Genus CECROPS, Leach.

Cecrops Latreillii, Leach.

Common on the gills of the sunfish (*Orthagoriscus mola*).

Genus ANCHORELLA, Cuvier.

Anchorella uncinata, O. F. Müller.

Abundant on the gills of cod and haddock.

T

Anchorella emarginata, Kröyer, Naturhist. Tidsskrift, Band i.
p. 287, tab. 3. fig. 7, *a–e*.

On the gills of the wolf fish (*Anarrhichas lupus*).   Mr.
Norman states that this is new to Britain.

Genus LERNÆA, L.

Lernæa branchialis, L.

Common on the gills of cod and haddock.

An *Ergasilus?* is frequent on the gills and other parts of
Doris tuberculata, D. Johnstoni, and occasionally on *Triopa
claviger.*

Order LOPHYROPODA.

Suborder OSTRACODA.

Fam. Cytheridæ.

Genus CYTHERE, O. F. Müller.

Cythere pellucida, Baird; G. S. Brady, Monogr. Brit. Ostracoda, Linn. Trans. xxvi. 2, p. 397, pl. 28. f. 22–26 & 28.

Abundant in shell-sand from the West Sands.

The following come from the same locality :—

Cythere albomaculata, Baird; Brady, *op. cit.* p. 402, pl. 28.
f. 33–39, pl. 39. f. 3.

Cythere lutea, O. F. Müller; Brady, *op. cit.* p. 395, pl. 28.
f. 47–56, pl. 39. f. 2.

Cythere villosa, G. O. Sars; Brady, *op. cit.* p. 411, pl. 29.
f. 28–32.

Cythere cuneiformis, Brady, *op. cit.* p. 404, pl. 31. f. 47–54.

Cythere viridis, O. F. Müller; Brady, *op. cit.* p. 397, pl. 28.
f. 40, 41, &c.   Also from deep water.

Cythere tuberculata, G. O. Sars; Brady, *op. cit.* p. 406, pl. 30.
f. 25–41.

Cythere concinna, Jones; Brady, *op. cit.* p. 408, pl. 26.
f. 28–33 &c.

Cythere finmarchica, G. O. Sars; Brady, *op. cit.* p. 410, pl. 31.
f. 9–13.

Genus CYTHERIDEA, Bosquet.

Cytheridea elongata, Brady, *op. cit.* p. 421, pl. 28. f. 13–16 &c.

Cytheridea papillosa, Bosquet; Brady, *op. cit.* p. 423, pl. 28.
f. 1–6 &c.

Genus LOXOCONCHA, G. O. Sars.

Loxoconcha tamarindus, Jones; Brady, *op. cit.* p. 435,
pl. 27. f. 45–48.

Occasionally in the débris of the fishing-boats.

Loxoconcha guttata, Norman; Brady, *op. cit.* p. 436, pl. 27.
f. 40–44.

In shell-débris from the West Sands.

Genus XESTOLEBERIS, G. O. Sars.

Xestoleberis aurantia, Baird; Brady, *op. cit.* p. 437, pl. 27.
f. 34–37 &c.

Abundant in tide-pools.

Genus CYTHEROPTERON, G. O. Sars.

Cytheropteron latissimum, Norman; Brady, *op. cit.* p. 448,
pl. 34. f. 26–30.

In the débris of the fishing-boats.

Genus CYTHERIDEIS, Jones.

Cytherideis subulata, Brady, *op. cit.* p. 454, pl. 35. f. 43–46.

In shell-débris from the West Sands.

Genus SCLEROCHILUS, G. O. Sars.

Sclerochilus contortus, Norman; Brady, *op. cit.* p. 455,
pl. 34. f. 5–10 &c.

Common in débris from deep water.

T 2

Genus PARADOXOSTOMA, Fischer.

Paradoxostoma variabile, Baird ; Brady, *op. cit.* p. 457,
pl. 35. f. 1–7 & 12–17.

Abundant in tide-pools and in deep water.

Paradoxostoma ensiforme, Brady, *op. cit.* p. 460, pl. 35.
f. 8–11.

In the débris of the fishing-boats.

Paradoxostoma flexuosum, Brady, *op. cit.* p. 461, pl. 35.
f. 30–34.

In the same locality.

Paradoxostoma arcuatum, Brady, *op. cit.* p. 461, pl. 35.
f. 37 & 38.

With the foregoing from deep water

Crustacean parasite of *Conchia
glauca*, from Lochmaddy.

Crustacean parasite of *Antedon*, from Lochmaddy.

Order AMPHIPODA.

Group **NORMALIA**.

Division *GAMMARINA*.   Subdivision VAGANTIA.

Tribe **Saltatoria**.

Fam. 1. **Orchestiidæ**.

Genus TALITRUS, Latreille.

Talitrus locusta, L.   Bate & Westwood, Brit. Sessile-eyed
			Crust. i. p. 16.

Abundant amongst the débris of seaweed and dead animals
of all kinds near high-water mark, and in burrows in the sand
even above the latter.

Genus HYALE, H. Rathke.

Hyale Nilssoni, H. Rathke ; B. & W. *op. cit.* i. p. 40 (as
			Allorchestes Nilssonii).

In small pools near high-water mark on the surface of the
bare rocks beyond the Maiden Rock, where almost the only
vegetation is borne on the backs of the limpets, and under
stones in littoral pools at the West Rocks.   Stomachs of the
cod and flounder.

Tribe **Natatoria**.

Fam. 2. **Gammaridæ**.

Subfamily *STEGOCEPHALIDES*.

Genus STENOTHOË, Dana (= *Probolium*, Costa ; *Montagua*,
			Bate & Westwood).

Stenothoë monoculoides, Mont. ; B. & W. *op. cit.* i. p. 54.

In débris of fishing-boats, not uncommon.   The dorsum
has rows of orange or reddish-orange specks, three distinct
rows on the broad plates of the anterior limbs, and other
isolated spots of the same hues ; eyes orange or reddish orange,
with small red dots posteriorly.   A variety also occurs.

Stenothoë marina, Bate; B. & W. *op. cit.* i. p. 58.
Frequent in débris of fishing-boats.   Ova green.

Stenothoë Alderi, Bate; B. & W. *op. cit.* i. p. 61.
With the foregoing, occasionally.

Stenothoë pollexiana, Bate; B. & W. *op. cit.* i. p. 64.
In the same locality.   Body barred with red; eyes red.

Stenothoë clypeata, Bate; B. & W. *op. cit.* ii. Supplement,
p. 499.
Occasionally in the débris of the fishing-boats.

Genus LYSIANASSA, M.-Edwards.
Lysianassa atlantica, M.-Edwards; B. & W. *op. cit.* i. p. 82.
Not uncommon in the stomach of the haddock.

Genus ANONYX, Kröyer.
Anonyx Holbüllii, Kröyer, = *A. denticulatus,* B. & W.
op. cit. i. p. 101.
Occasionally after storms on the West Sands, and in the
stomachs of cod and haddock.

Genus ACIDOSTOMA, Lilljeborg.
Acidostoma obesum, Bate; B. & W. *op. cit.* i. p. 98.
Occasionally at the East Rocks.

Genus CALLISOMA, Costa.
Callisoma crenata, Bate; B. & W. *op. cit.* i. p. 120.
In the stomach of a haddock.

Subfamily *AMPELISCIDES.*

Genus AMPELISCA, Kröyer.
Ampelisca carinata, Bruzelius; B. & W. *op. cit.* i. p. 127
(as *A. Gaimardii*).
Abundant in the stomachs of cod and haddock.

Ampelisca Belliana, Bate (= *A. macrocephala,* Lilljeborg ?) ;
B. & W. *op. cit.* i. p. 135.

Common in the stomachs of the cod, haddock, skate, and
flounder, and dredged off the East Rocks. Nothing else is
found in the distended stomachs of some haddocks except
masses of this species ; or they may be accompanied by green
pea-urchins, tubes of *Terebella,* fragments of *Ophiocoma,* and
sea-mice. In multitudes on the West Sands after some
storms.

Genus AMPHILOCHUS, Bate.

Amphilochus manudens, Bate; B. & W. *op. cit.* i. p. 180.

Occasionally in the débris of the fishing-boats. Eyes bright
red; body purplish brown, speckled with dark granules; the
tips of the antennæ have the same purplish hue.

Genus IPHIMEDIA, H. Rathke.

Iphimedia obesa, H. Rathke; B. & W. *op. cit.* i. p. 219.

Not uncommon in pools at the East Rocks, and in débris
of the fishing-boats. The brownish-red markings of the
young specimens form a double row on the posterior segments.

Subfamily *GAMMARIDES.*

Genus DEXAMINE, Leach.

Dexamine spinosa, Mont. ; B. & W. *op. cit.* i. p. 237.

Abundant in pools near low-water mark at the East Rocks,
and in the stomach of the cod. Eyes white. Most have a
straw-coloured body, very prettily mottled with brownish-red
patches and many minute white specks; the antennæ are
beautifully barred with white and brown.

Genus ATYLUS, Leach.

Atylus Swammerdamii, M.-Edwards; B. & W. *op. cit.* i. p. 246.

Occasionally in rock-pools at the pier, or clinging in hundreds
to the seaweeds in the laminarian region off the West Rocks ;
abundant on the beach after storms, and in the stomach of the

cod. Translucent and slightly yellowish, with three brownish-
red spots along the dorsum, and a small one above the eyes ;
the latter are pinkish brown ; the elongated heart pulsates
very evidently on the dorsum.

Atylus bispinosus, Bate ; B. & W. *op. cit.* i. p. 250.

In the débris of the fishing-boats, under stones at the Pier
rocks, and on the West Sands after storms. Eyes occasionally
reddish. Most of the body and appendages are speckled with
small black dots ; many have specks of a carmine hue behind
the eyes.

Genus PHERUSA, Leach.

Pherusa bicuspis, Kröyer ; B. & W. *op. cit.* i. p. 253.

In the débris of the fishing-boats, and in swarms in the
fine pools near high-water mark beyond the Rock and Spindle.

Genus CALLIOPIUS (Leach), Lilljeborg.

Calliopius læviusculus, Kröyer ; B. & W. *op. cit.* i. p. 259.
Occasionally in pools near low water at the East Rocks.

Calliopius Ossiani, Bate ; B. & W. *op. cit.* i. p. 261.
Frequent in the fishing-boats.

Calliopius grandoculis, Bate ; B. & W. *op. cit.* i. p. 265.

In the same locality, and not uncommon in the rock-pools.
Many show a decided brownish bar from the eyes along the
dorsal ridge ; and sometimes small reddish specks are present.
A. Boeck includes this form under *C. læviusculus* [*].

Calliopius bidentatus (n. sp.), Norman, Nat. Hist. Trans.
Northumb. & Durham, vol. i. 1865, p. 24.

This species is frequently dredged off the Harbour and the
East Rocks, as well as in the deeper water outside the bay,
and found on the West Sands after storms. Mr. Norman
states that it is not uncommon all along the east coast.

[*] 'Crustacea Amphipoda borealia et arctica,' p. 117.

The body is about two fifths of an inch long, of a pale straw-colour, tinted with brownish at the joints and the bases of the limbs. Superior antennæ twice as long as the inferior, beautifully banded with red. Eyes irregularly rounded, brownish red or pale brick-red. The first and second gnathopods are nearly equal (the second, however, being larger) and similar in structure. Hand almond-shaped, the palm furnished with a series of very distinct stout spines, and a row of smaller spines reaching the base of the finger ; the latter is long, boldly curved, and regularly divided on the concave side. The first and second pleopods have spines, that of the former, however, being sometimes indistinct. A very characteristic convexity occurs at the junction of the third and fourth pleopods ; and the dorsal margin of the latter is concave.

Genus LEUCOTHOË, Leach.

Leucothoë spinicarpa, Abildgaard ; B. & W. *op. cit.* i. p. 271 (as *L. articulosa*).

Occasionally in pools at the East Rocks, and on the West Sands after storms.

Genus AORA, Kröyer.

Aora gracilis, Bate ; B. & W. *op. cit.* i. p. 281.

Not uncommon in the débris of the fishing-boats. One had a spike beneath the second pair of gnathopods.

Genus MICRODEUTEROPUS, Costa.

Microdeuteropus Websteri, Bate ; B. & W. *op. cit.* i. p. 291.

In the stomach of a haddock, and in débris of the fishing-boats. Body of a straw-colour, the antennæ having lighter and darker bands of the same hue ; eyes round, black.

Genus BATHYPOREIA, Lindström.

Bathyporeia pilosa, Lindström ; B. & W. *op. cit.* i. p. 304.

Common off the East Rocks in the laminarian region.

U

Bathyporeia Robertsoni, Bate; B. & W. *op. cit.* i. p. 309.
Occasionally in pools at the East Rocks. The eyes in the
examples were large, nearly meeting in the middle line.

Genus MELITA, Leach.

Melita palmata, Mont.; B. & W. *op. cit.* i. p. 337.
In the débris of the fishing-boats; not common. The body
is yellowish or straw-colour, with pale brownish antennæ
marked at the joints with pale rings; eyes dark brown or
black, with whitish specks.

Melita obtusata, Mont.; B. & W. *op. cit.* i. p. 341.
From the fishing-boats; not uncommon.

Genus GAMMAROPSIS, Lilljeborg.

Gammaropsis erythrophthalmus, Lilljeborg; B. & W. *op. cit.* i.
p. 354.
From the fishing-boats; not rare.

Genus AMATHILLA, H. Rathke.

Amathilla Sabini, Leach; B. & W. *op. cit.* i. p. 361.
A single example in the stomach of a haddock.

Genus GAMMARUS, Fab.

Gammarus marinus, Leach; B. & W. *op. cit.* i. p. 370.
In the stomach of a cod, and occasionally off the East Rocks
in a few fathoms.

Gammarus locusta, L.; B. & W. *op. cit.* i. p. 378.
In swarms below the flat stones on sand between tide-
marks and in the laminarian region. It swims a considerable time in putrid water. Occurs frequently in the stomachs
of cod and haddock.

Genus HEISCLADIUS, B. & W.

Heiscladius longicaudatus, B. & W. *op. cit.* i. p. 412.
In the fishing-boats ; rare.

Subdivision DOMICOLA.

Fam. **Corophiidæ.**

Subfamily *PODOCERIDES.*

Genus AMPHITHOË, Leach.

Amphithoë rubricata, Mont. ; B. & W. *op. cit.* i. p. 418.
In the débris of the fishing-boats.

Amphithoë podoceroides, H. Rathke ; B. & W. *op. cit.* i. p. 422
(as *A. littorina*).

Common in the laminarian region, and under stones between
tide-marks, where it constructs tubes or nests. Most of the
fine specimens have the hand of the second pair defined by a
distinct tooth, as Rathke and Dr. Johnston state.

Genus PODOCERUS, Leach.

Podocerus falcatus, Mont. ; B. & W. *op. cit.* i. pp. 436 & 447 (as
P. pulchellus and *P. pelagicus*).

In rock-pools on *Ceramium rubrum* at the Pier, in the
laminarian region beyond, in the stomachs of flounders, and
in the fishing-boats. Sometimes gaudily tinted with reddish
brown and white, and with red bars on the inferior antennæ.

Podocerus variegatus, Leach; B. & W. *op. cit.* i. p. 439, & p. 442
(as *P. capillatus*).

Not uncommon in pools near low-water mark at the
East Rocks.

Genus CERAPUS, Say.

Cerapus difformis, M.-Edwards; B. & W. *op. cit.* i.
p. 457.

Common in deep water. The straw-coloured body is marked
with dark grains; and the superior antennæ have the basal
third of the second and third segments tinted of a crimson
hue, the flagellum being similarly coloured for its proximal
half; the eyes have black centres and, as usual, a pale
margin.

Genus SIPHONŒCETUS, Kröyer.

Siphonœcetus typicus, Kröyer; B. & W. *op. cit.* i. p. 465.

Abundant in the laminarian region in 3 to 6 fathoms off
the East Rocks, where it constructs nests on the inner surface
of bivalve shells. *S. Whitei*, Gosse, is probably the female of
this species.

Genus NÆNIA, Bate.

Nænia tuberculosa, Bate; B. & W. *op. cit.* i. p. 472.
Occasionally in the débris of the fishing-boats.

Nænia rimopalmata, Bate; B. & W. *op. cit.* i. p. 474.
With the former.

Nænia excavata, Bate; B. & W. *op. cit.* i. p. 476.
Common in the same débris from the coralline ground.

Subfamily *COROPHIIDÆ*.

Genus COROPHIUM, Latreille.

Corophium grossipes, L.; B. & W. *op. cit.* i. p. 493 (as *C. longicornis*).

Abundant in the brackish pools near the mouth of the Eden,
and occasionally at the West Rocks. It is common in July:
swims excellently on its back.

Division *HYPERINA.*

Fam. **Hyperiidæ.**

Genus HYPERIA, Latreille.

Hyperia medusarum, O. F. Müller; B. & W. *op. cit.* ii. p. 12
(as *H. galba*).

Common in the cavity of *Aurelia aurita*; each medusa had six or eight large examples. The *Lestrigonus Kinahani,* Bate, is a sexual variety (male). Some large specimens are found swimming freely on the surface of the water.

Hyperia oblivia, B. & W. *op. cit.* ii. p. 17.

In a tide-pool on the West Sands after a storm.

Group **ABERRANTIA.**

Fam. **Caprellidæ.**

Genus ÆGINA, Kröyer.

Ægina phasma, Mont.; B. & W. *op. cit.* ii. p. 45.

Abundant in the débris of the fishing-boats.

Genus CAPRELLA, Lamarck.

Caprella linearis, L.; B. & W. *op. cit.* ii. p. 52.

Plentiful in the same locality.

Caprella lobata, O. F. Müller; B. & W. *op. cit.* ii. p. 57.

Frequent in the fishing-boats.

Caprella tuberculata, Guérin; B. & W. *op. cit.* ii. p. 68.

Common on *Ceramium rubrum* in rock-pools, and in the stomachs of cod and haddock.

Caprella hystrix, Bate; B. & W. *op. cit.* ii. p. 63.

Not uncommon in the fishing-boats. The Rev. A. M. Norman does not think this is the *C. hystrix* of Kröyer, but rather the *C. septentrionalis* of that author.

Order ISOPODA.

Group ABERRANTIA.

Tribe *VAGANTIA*.

Genus ANCEUS, Risso.

Anceus maxillaris, Mont. ; B. & W. *op. cit.* ii. p. 187.
Not uncommon in the débris from the coralline ground.

Division AQUISPIRANTIA.

Tribe *PARASITICA*.

Fam. Bopyridæ.

Genus PHRYXUS, H. Rathke.

Phryxus paguri, H. Rathke ; B. & W. *op. cit.* ii. p. 240.
Occasionally on *Pagurus bernhardus*.

Fam. Ægidæ.

Genus CIROLANA, Leach.

Cirolana spinipes, M.-Edw. ; B. & W. *op. cit.* ii. p. 299.
A large specimen occurred in the stomach of a haddock.

Genus EURYDICE, Leach.

Eurydice pulchra, Leach ; B. & W. *op. cit.* ii. p. 310.
Abundant on the surface of the sea off the East Rocks
in autumn, and in the stomachs of cod and haddock.

Tribe *LIBERATICA*.

Fam. Asellidæ.

Genus JÆRA, Leach.

Jæra Nordmanni, H. Rathke ; B. & W. *op. cit.* ii. p. 320.
Common under stones near high-water mark at the East
Rocks.

Genus JANIRA, Leach.

Janira maculosa, Leach; B. & W. *op. cit.* ii. p. 338.

Frequent on shells and *Filigrana* from the coralline ground, and under stones in pools near low water at the East and other rocks. This species has many of the habits of *Idotea*.

Genus LIMNORIA, Leach.

Limnoria lignorum, Rathke; B. & W. *op. cit.* ii. p. 351.

Abundant in the stakes for the salmon-nets on the West Sands, and in wood elsewhere.

Fam. Arcturidæ.

Genus ARCTURUS, Latreille.

Arcturus longicornis, Sowerby; B. & W. *op. cit.* ii. p. 365.

Common in the stomachs of cod, haddock, and flounders.

Arcturus gracilis, H. Goodsir; B. & W. *op. cit.* ii. p. 373.

Abundant in débris from the coralline ground and in the stomachs of haddocks.

Fam. Idoteidæ.

Genus IDOTEA, Fab.

Idotea tricuspidata, Desmarest; B. & W. *op. cit.* ii. p. 379.

Frequent near low water in the laminarian region, and in the stomachs of all the common fishes.

Idotea linearis, Pennant; B. & W. *op. cit.* ii. p. 388.

Common in 3 or 4 fathoms on sand near the bar of the Eden, in the trawlers' boats, and in the stomachs of the common fishes. They are active swimmers.

Messrs. Bate and Westwood state that I sent *Cymodocea truncata*, Mont., from St. Andrews; but this is doubtful. The specimens probably came from the Outer Hebrides.

Division **AERISPIRANTIA**.

Fam. **Oniscidæ**.

Genus LYGIA, Fab.

Lygia oceanica, L.; B. & W. *op. cit.* ii. p. 444.
Abundant at the margin of high water at the East Rocks.

A specimen of *Porcellio scaber* occurred in the stomach of a cod.

Order CUMACEÆ.

Fam. **Diastylidæ**.

Genus DIASTYLIS, Say.

Diastylis Rathkii, Kröyer.
Common off the East Rocks in 3 to 4 fathoms, and in the stomach of the cod, haddock, and flounder.

Order PODOPHTHALMATA.

The stalk-eyed Crustacea of St. Andrews are chiefly northern in type; and though the species are not numerous, many
are very plentifully represented. The most important forms
here, as elsewhere, are the edible crab and the lobster. Both
are caught in considerable numbers along the border of the
rocks by means of the ordinary crab-pots, which are generally
baited with fragments of grey gurnards and other fishes of
little value. The most successful ground is off the East
Rocks, though a very large lobster in the Museum of the
University was procured to the north of the West Rocks.
Some of the fishermen have an idea that if a lobster enters
a trap first, none of the edible crabs will venture beside it,
whereas a lobster will invade the crab-pot though a dozen of
the former are already there. Constant attacks seem to have
diminished the numbers of both species, and especially of the
lobster. I have never seen any of the latter between tide-
marks; but young edible crabs are common under ledges and
stones, and even in the sand at low water, their presence in
the latter being recognized by a depression. The common
shore-crab occurs everywhere along the rocky border, both
between tide-marks and in the laminarian region. This ubiquitous species lurks in the retired apertures and clefts amongst
and under the rocks, especially where these have a bottom of
soft sand or dark mud. In this it buries itself so as to retain
moisture in the gills, while the anterior part of the carapace is
uncovered, probably for quiet observation. In these situations
it quite understands an attempt to capture it; and there are
few examples, if any, in which, by seizing the crooked iron
with its chelæ, it has allowed itself to be drawn out. On the
contrary, it endeavours to escape with much effort and considerable agility. Even when quite invisible its presence may be
detected by striking the rock, when the grating of the carapace
is heard as the animal retreats. It is often to be found in
positions which seem any thing but comfortable—amongst
blackened and putrefying animal remains, in muddy and
odoriferous pools tenanted by none except itself. In these

x

circumstances the body is coated with mud, which fills up the
irregularities of its conformation, and loads the abdominal
feet and hairs ; yet the crab is vigorous and healthy, and out-
lives sanitary apprehensions.

Under almost every stone within reach of the tide young
specimens occur. At low water the full-grown crabs seek the
hiding-places just mentioned, or shade themselves under the
blades of the seaweeds in the rock-pools. Occasionally one is
found adhering to the soft body of a moulting brother and,
cannibal-like, devouring the branchiæ, new carapace, and other
soft organs with savage pertinacity, while the old shell has
not quite fallen from its victim. Moulting shore-crabs are
generally found alone, as if aware of their helplessness,
and dreading, with some degree of justice, the voracity of
enemies and even unscrupulous relations. Very slight injury
kills them in this condition ; and of course, for a time, they are
incapable of defending themselves from even weak assailants.

The shore-crab is found in pools at the East Rocks where no
other marine articulate of the same class occurs, and the water
cannot but be brackish, since the pools are not filled by ordinary
tides, and fresh streams from the crags flow in the neighbour-
hood. In these resorts the colour of the crab is not so pretty,
being of a muddy green with pale limbs; and the specimens
in the highest pools are generally small. It is not surprising,
however, to find them in such places, after watching their
activity in the innumerable brackish lakes of the Outer
Hebrides, and their evident comfort in perambulating the
muddy flats even where streams of fresh water abound.

On land, *Carcinus mœnas* is, perhaps, the most active British
crab, especially in regard to offence, defence, and escape. It
scrambles over the rugged rocks with astonishing speed, while
defending itself with its uplifted chelæ ; and so fierce is it in
attack, that, having once seized an object with the latter, the
spasmodic effort is sometimes so great that the limb separates
from the trunk at the base. The males frequently engage in
combat; and a fatal issue would more frequently ensue, were it
not for the provision whereby hæmorrhage is speedily arrested
and the lost portion repaired or reproduced. Few specimens,
indeed, are quite free from injury. Some have recently repaired
wounds of the carapace (Plate IX. fig. 12) ; others have lost

an eye, an antenna, or one or more limbs. They surpass most marine animals in their powers of enduring life at a distance from sea-water, and may easily be kept for several weeks in a botanic vasculum.

The shore-crab is strictly carnivorous and, as already mentioned, even relishes its fellows. It is a curious feature in its history that it suffers serious annoyance and injury from the young of the common mussel, which plant themselves in its orbits (Plate IX. fig. 11), in the sockets of the internal antennæ, in the branchial chambers, and under the tail (Plate IX. fig. 10)—in the former case often destroying both eyes. It feeds with avidity on the mussel in its adult state; so that here is an instance of a helpless young form avenging the destruction of the mature. The shore-crab, again, is devoured by many fishes: thus in the stomach of a *Cottus bubalis* I have found five or six specimens, two entire and upwards of 2 inches across the carapace. The *Cottus*, however, unfortunately came in the way of a large frogfish, which found a place for it in its capacious stomach, though nine full-grown flounders were already present. In many parts of Britain and the continent the shore-crab is used as food by man (and this is a safe-enough practice so long as it is well boiled, internal parasites being abundant); but at St. Andrews it is only employed occasionally for bait.

Myriads of the young of this species in the zoën-stage occur at the surface of the bay in autumn, and may easily be kept alive, so as to show the subsequent stages of development.

Besides those already mentioned, many of the other forms are very common, such as *Stenorhynchus rostratus*, *Inachus*, *Hyas*, *Portumnus variegatus*, the *Portuni*, *Pinnotheres*, *Ebalia*, and *Nephrops* in deep water, *Porcellana*, the *Paguri*, *Galathea*, and *Crangon* between tide-marks, and in both regions *Hippolyte*, *Pandalus*, and *Palæmon*. In deep water swarms of *Hyas coarctatus* for the most part take the place of *H. araneus*. As a littoral form *Palæmon squilla* is local, but, in company with *Pandalus annulicornis*, it is abundant in deep water. The common shrimp is seldom captured by man for food. *Portumnus variegatus* is often the only form visible on the West Sands, and is very plentiful. The rarer forms are *Eurynome*, *Piri-*

mela, Lithodes, Gebia deltura, Hippolyte spinus, and *Doryphorus Gordoni.*

In contrast with the fauna of St. Andrews, we have in the mild sea of the west of Scotland the fine velvet crabs (*Portunus puber*) amongst the seaweeds between tide-marks. The common lobster is also much more abundant, though the wholesale fishing has of late years told severely on this crustacean, even on the most remote shores of the Outer Hebrides—as, for instance, off the rocks of Haskeir near the north-west point of North Uist, where the frequent inroads of the fishermen with their lobster-pots and floats have rendered even the seals less frequent in their accustomed haunts. *Xantho, Munida,* and the rarer species of *Crangon* and *Hippolyte* are also absent from St. Andrews. In the south of Britain, again, are the splendid spiny lobsters off the rocky shores, velvet crabs, *Pirimela,* and *Ebalia* under stones between tide-marks, *Alpheus ruber* and *Pagurus cuanensis* in littoral pools, *Pilumnus* in the crevices of the tidal rocks, *Pagurus Prideauxii* with the beautiful *Adamsia* adherent to its protecting shell, *Maia* *, *Dromia,* and *Polybius.* In the northern waters swarms of the hardy *Portunus pusillus, P. tuberculatus, Pagurus pubescens,* and *Pandalus brevirostris* are characteristic, besides the rarer *Pagurus tricarinatus, Crangon serratus,* and *Sabinea septemcarinata.*

I am indebted to the Rev. A. M. Norman for kind assistance with several species of Palæmonidæ and Galatheidæ.

Suborder STOMAPODA.

Fam. Mysidæ.

Genus MYSIS, Latreille.

Mysis flexuosa, O. F. Müller; Bell, Brit. Crust. p. 336
(as *M. chamæleon*).

Very abundant in rock-pools.

---

* It was recently stated in ' Land and Water' that *Maia squinado* had been procured near the Bell Rock; but, by the kindness of Mr. F. Buckland, who forwarded the specimen, I am enabled to observe that it was only *Lithodes maia.*

Mysis vulgaris, J. V. Thompson ; Bell, *op. cit.* p. 339.
Occasionally with the former in rock-pools; much less common.

Mysis Griffithsiæ, Bell, *op. cit.* p. 342.
Not uncommon in rock-pools, and occasionally thrown on the West Sands in multitudes after storms.

Suborder **DECAPODA.**

Tribe *MACRURA.*

Fam. **Palæmonidæ.**

Genus PALÆMON, Fab.

Palæmon squilla, L. ; Bell, *op. cit.* p. 305.
Common in pools beyond the Rock and Spindle and in the stomachs of cod.

Genus PANDALUS, Leach.

Pandalus annulicornis, Leach ; Bell, *op. cit.* p. 297.
Abundant from the laminarian region to deep water, and also in the stomachs of cod and haddock.

Genus HIPPOLYTE, Leach.

Hippolyte varians, Leach ; Bell, *op. cit.* p. 286.
Frequent in rock-pools and ranging thence to deep water; stomachs of haddock.

Hippolyte pusiola, Kröyer, Monogr. af Slægten Hippolytes Nordiske Arter, p. 319, pl. 3. f. 69–73 (*fide* Rev. A. M. Norman).
Occasionally from the coralline ground amongst shells and stones, and in pools at the East Rocks.

Hippolyte securifrons, Norman, Tyneside Nat. Field-Club
Trans. vol. v. (1863), pl. 12. figs. 1–7.
Occasionally in the stomach of the flounder.

Hippolyte spinus, Sowerby ; Bell, *op. cit.* p. 284.
Occasionally in the stomach of the haddock.

Genus DORYPHORUS, Bate.

Doryphorus Gordoni, Bate, Nat. Hist. Review, vol. v. (1858),
p. 51.
Under a large stone in a pool near low water at the East
Rocks.   Rare.

Fam. Crangonidæ.

Genus CRANGON, Fab.

Crangon vulgaris, Fab. ; Bell, *op. cit.* p. 256.
Abundant off the West Sands and in sandy tide-pools, as
well as on the beach after storms.

Fam. Astacidæ.

Genus NEPHROPS, Leach.

Nephrops norvegicus, L.; Bell, *op. cit.* p. 251.
Common in deep water and in the stomachs of cod.

Genus HOMARUS, M.-Edwards.

Homarus gammarus, L. ; Bell, *op. cit.* p. 242.
Common in the laminarian region.

Fam. Thalassinidæ.

Genus GEBIA, Leach.

Gebia deltura, Leach ; Bell, *op. cit.* p. 225.
Occasionally in the stomachs of cod and haddock.

Tribe *Anomura*.

Fam. **Galatheidæ.**

Genus GALATHEA, Fab.

Galathea strigosa, L.; Bell, *op. cit.* p. 200.

Not uncommon in deep water and in the stomachs of cod and haddock.

Galathea squamifera, Mont.; Bell, *op. cit.* p. 197.

Very common under stones near low water, especially in pools and runlets; occasionally in the stomachs of cod.

Galathea dispersa, Bate, Proceed. Linn. Soc., Zool. vol. iii. p. 3.

Abundant in deep water, and in the stomachs of the cod, haddock, and flounder.

Fam. **Paguridæ.**

Genus PAGURUS, Fab.

Pagurus bernhardus, L.; Bell, *op. cit.* p. 171.

Everywhere abundant between tide-marks and in deep water. A young specimen was lodged inside a fragment of a stalk of wheat.

This species has nine or ten branchiæ on each side, besides a rudimentary organ at the base of the first pair of foot-jaws. The latter have no branchial whips, and differ considerably from those of the Brachyura.

The first pair of foot-jaws have the inner division very much elongated, almost antenniform, and bordered with long hairs, while the external portion is small. In the next pair the inner division more closely agrees with the external in length, and the whole is not very different from the same part in *Carcinus mœnas* minus the whip and branchia. The third pair is shorn of its whip and large flap, and has the middle segment * repre-

* Corresponding to *d*, fig. 3, Trans. Linn. Soc. vol. xxiv. p. 86.

sented by a narrow pedicle. The fourth pair has a narrow shield turned over at the free edge, and, instead of the two narrow spikes below, there is a flattened organ which forks into a narrow and a broad flap at the tip. The fifth pair has its inner division broad and flattened, and its outer small, but widened at the tip; the median division has a very regular arrangement of bristles at its tip, which points or slopes inwards.

The parasitic *Peltogaster paguri* frequently occurs on the abdomen.

Pagurus cuanensis, Thompson ; Bell, *op. cit.* p. 178.
Occasionally from deep water.

Pagurus ulidianus, Thompson (?); Bell, *op. cit.* p. 180.
St. Andrews Museum. I cannot speak with certainty of this form.

Pagurus lævis, Thompson ; Bell, *op. cit.* p. 184. '
Occasionally in the stomach of the haddock.

Fam. Porcellanidæ.

Genus PORCELLANA, Lamarck.

Porcellana platycheles, Penn. ; Bell, *op. cit.* p. 190.

Abundant under stones between tide-marks, especially in runlets, and on muddy ground. A group of young forms of some size may sometimes be seen in company with their parents.

The first pair of foot-jaws have their two terminal segments furnished with the longest hairs (proportionally) yet met with in the local forms. The hairs have a double row of spikes, diminishing towards base and tip, and ceasing before arriving at the end of the hair, which has very fine linear serrations. The external division has a powerful triangular, and somewhat tapering, lower segment, and a delicate appendage fringed with a brush of spiked hairs at the tip. The second pair has the external division much flattened, lanceolate, and with hairs having spiked bases and serrated tips on the outer edge ; the hairs also occur generally along the inner margin, and are frequently sheathed in mud and particles of all kinds. The third pair consists of three portions furnished with long branched . hairs. The fourth pair has the large flat shield surrounded with branched hairs ; next is a curved tapering portion with bristles having short spikes towards the tip ; then come a series of flattened organs with truncate tips covered with spiked hairs. The fifth pair has three divisions—an inner irregular portion with hairs shortly branched on its free edge, a middle and somewhat club-shaped piece with rather stiff serrated hairs scantily spiked at the base, and a curiously curved and rather slender inner portion with about half a dozen finely serrated hairs on one side of its tip.

The hairs on the outer border of the chelæ are densely plumose ; and hence it is exceedingly difficult to clean them from mud and sand for the cabinet.

Porcellana longicornis, L.; Bell, *op. cit.* p. 193.

As common as the former, in similar, though not muddy, situations. The embryos are found in the ova in August ; and many young occur under stones in November and December.

Fam. Lithodidæ.

Genus LITHODES, Latr.

Lithodes maia, L.; Bell, *op. cit.* p. 165.

Not uncommon in deep water, whence it is brought by the fishing-boats.

Tribe *Brachyura.*

Fam. Leucosiadæ.

Genus EBALIA, Leach.

Ebalia tuberosa, Penn.; Bell, *op. cit.* p. 141.

Not uncommon in the stomachs of cod, and occasionally from deep water.

Ebalia Cranchii, Leach; Bell, *op. cit.* p. 148. .

Occasionally in the stomach of the haddock.

Fam. Maiidæ.

Genus INACHUS, Fab.

Inachus dorsettensis, Penn.; Bell, *op. cit.* p. 13.

Not uncommon in the stomach of the cod.

Inachus dorhynchus, Leach; Bell, *op. cit.* p. 16.

Occasionally under stones near low-water mark. In the stomach of one were fragments of *Ulea,* and in another the débris of a large sessile-eyed crustacean. The hairs on this species are shaped like the horn of the chamois; and some have a slight enlargement at the base.

Genus HYAS, Leach.

Hyas araneus, L.; Bell, *op. cit.* p. 31.

Abundant under ledges in rock-pools, cast ashore on the West Sands after storms, in the crab-pots, and in the stomach of the cod.

This species has eight branchial processes on each side—four lateral, two anterior, and one to each of the first two pairs of foot-jaws. Their structure resembles that described in *Carcinus mænas.* The ova apparently of a small leech (*Ponto-*

bulella) are often found attached to the walls of the branchial chamber.

The number and variety of parasitic growths, both vegetable and animal, on the carapace of this form are remarkable. *Balani* of two species cover the back almost with a continuous rugose pile, adhering to the limbs, the abdomen, the foot-jaws, or each other. Coils of *Serpulæ* and hard sandy tubes of *Sabellaria* interlace with these and fill up the depressions, and, with the former, occur on the tip of the abdomen as well as in less mobile situations. Fine tufts of *Sertularia pumila* and *Crisia eburnea* adorn the surface of the carapace in others, or the parasitic algæ thereon; while *Halichondria panicea* forms a thick rugged crust, from which *Balani, Serpulæ, Anomiæ*, zoophytes, and seaweeds emerge. Even the sockets of the eyes are invaded by the sponge. Moreover young examples are not unfrequently clothed with thick tufts of *Obelia geniculata*. It would appear that it is not always on attaining full growth that moulting ceases for considerable intervals, since small specimens are found as completely covered with parasitic growths. In the rock-pools the carapace often forms a moving forest of seaweeds; and in such specimens the shell is frequently fragile, so that the extraneous covering may be of use for protection, or else had grown with unusual rapidity, even before the carapace became fully consolidated.

One old example had the internal antennæ quite fixed by a hard sand-tube of *Sabellaria*; and the young of the common mussel are occasionally found in the cavities for the eyes.

In the young females the genital apertures are small, and the abdomen less developed; while in the adult the latter becomes hypertrophied, hollowed out on its ventral surface by the bending downwards of the outer edges, and touches the bases of the legs on each side.

Hyas coarctatus, Leach; Bell, *op. cit.* p. 35.

Common in deep water, and procured in hundreds amongst the coralline débris in the fishing-boats; frequent in the stomachs of cod, haddock, and flounders.

Fam. Leptopodiadæ.

Genus STENORHYNCHUS, Lam.

Stenorhynchus rostratus, L.; Bell, *op. cit.* p. 2
(as *S. phalangium*).

Abundant in the coralline region, in the stomachs of cod
and haddock, and occasionally under stones at low water.
Fragments of sessile-eyed Crustacea and sand occurred in the
stomachs of those examined.   Males greatly preponderate.

Fam. Parthenopidæ.

Genus EURYNOME, Leach.

Eurynome aspera, Penn.; Bell, *op. cit.* p. 46.

A few specimens were procured from the coralline ground.
Rare.

Fam. Canceridæ.

Genus CANCER, L.

Cancer pagurus, L.; Bell, *op. cit.* p. 59.

Abundant all round the rocky border in the laminarian
region, and frequent between tide-marks.   In the stomach of
this species are many curious parasites, such as *Tetrarhynchus*
(Plate VII. figs. 16 & 17) and *Echinorhynchus*, probably derived from its food.   Sections of the carapace show internally
tubular processes, apparently connected with the hairs (Plate IX.
fig. 9).

Genus PIRIMELA, Leach.

Pirimela denticulata, Mont.; Bell, *op. cit.* p. 72.
Occasionally from deep water.   Rare.

Fam. Portunidæ.

Genus PORTUNUS, Leach.

Portunus depurator, L.; Bell, *op. cit.* p. 101.

Dredged occasionally off the West Rocks on a sandy bottom,
cast ashore by storms, or found in the stomach of the cod.

Portunus marmoreus, Leach ; Bell, *op. cit.* p. 105.
On the West Sands after storms. Rather rare.

Portunus holsatus, Fab. ; Bell, *op. cit.* p. 109.
Not uncommon in the stomachs of cod and haddock. *Sacculina* occurs on this species occasionally (Plate IX. figs. 14 & 15).

Portunus pusillus, Leach ; Bell, *op. cit.* p. 112.
Occasionally from deep water, and rather common in the stomachs of the haddock and flounder.

Genus PORTUMNUS, Leach.
Portumnus variegatus, Leach ; Bell, *op. cit.* p. 85.
Abundant on the sandy ground off the West Sands.

Genus CARCINUS, Leach.
Carcinus mænas, L. ; Bell, *op. cit.* p. 76.
[Plate V. figs. 7 & 8, and Plate IX. figs. 10-13.]

Everywhere abundant between tide-marks and in the laminarian region. Occasionally used as bait. Swarms in the zoëa-stage occur in autumn at the surface of the water in the bay; they are almost invisible with the exception of the greenish-blue eyes.

This crustacean has nine branchiæ :—the first rudimentary, and attached to the horizontal portion of the first pair of foot-jaws (Plate V. fig. 7) ; the succeeding, rather long and delicate organs, fixed to the second pair of foot-jaws on opposite sides of the horizontal portion ; while six are attached to the body of the animal (four being prominent) as in allied forms. The flabellum of the first pair passes between the four prominent and larger branchiæ and the apodematous region, so as to sweep their inner surface ; while the organ of the second pair goes between the same portion of the shell and the fifth and sixth branchiæ (counting from behind), and may also affect the exposed surface of the seventh, which lies in the groove anteriorly. The long and finely curved flabellum of the third pair of foot-jaws curves externally, so as to brush all the seven.

The great development of this organ, its central calcareous bow, and long hairs are thus explained. The branchial laminæ are ranged with their edges to the afferent current, which crosses the organs at right angles to their long axes, and so impinges between the plates. The action of the broad shield of the fourth pair of foot-jaws, again, affects the ingoing stream, and plays upon the large flat surface at the base of the flabellum of the third pair. It would thus tend to spread out the long hairs of the latter, and direct the current upwards over the branchial laminæ. The fifth pair as a whole would seem to be connected with the buccal rather than the respiratory apparatus; for the curiously twisted portion (c, fig. 6, Trans. Linn. Soc. vol. xxiv. p. 88) is nicely adapted to the deep anterior notch of the mandible, while the curved portion (a) enters the mouth above the chitinous tissue filling up the posterior notch of the mandible. The tuft of long hairs (e, loc. cit.), however, may render some assistance to the branchial portion of the fourth pair of foot-jaws in contact with it.

The appendage of the mandible (a, fig. 7, op. cit.) seems to have a considerable influence in the prehension and direction of the food between the maxillæ; it has lateral motion as well as flexion and extension. The flexible process filling up the gap in the underpart of the maxilla, and connected with the lip beneath the latter, would seem to prevent the escape of particles in biting and deglutition. It is attached to a firm horny basis, which has free horizontal, but little or no vertical motion, except when greatly extended.

In females bearing eggs the muscles on the external or under surface of the intestinal tract greatly increase in size at the junction of the abdomen with the cephalothorax. In males and females without ova the exterior of the gut is sparingly supplied with such tissue.

This crab affords a good example of the " commensalisme " of Prof. van Beneden. Nemertes carcinophila abounds on the hairs bearing ova; and the young of the edible mussel and other adventitious growths are common, besides Sacculina (Plate IX. fig. 13), and Trematode larvæ in the liver and other parts. Various abnormalities from injury also occur. The colours of the males are often remarkably bright, both on the upper and under surfaces of the carapace (Plate V. fig. 8).

Fam. **Corystidæ**.

Genus ATELECYCLUS, Leach.

Atelecyclus septemdentatus, Mont.; Bell, *op. cit.* p. 153.
Frequent in the stomachs of cod.

Genus CORYSTES, Latreille.

Corystes cassivelaunus, Penn.; Bell, *op. cit.* p. 159.
Common on the West Sands after severe storms.

Fam. **Pinnotheridæ**.

Genus PINNOTHERES, Latreille.

Pinnotheres pisum, L.; Bell, *op. cit.* p. 121.
Frequent in *Mytilus modiolus*.

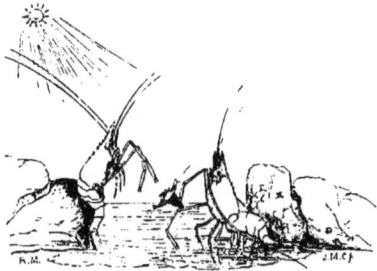

Subkingdom *VERTEBRATA.*

Class PISCES.

Contrasted with the extreme shores of Britain the fish-fauna of St. Andrews bay exhibits certain interesting differences, though of course its features are common to many other parts of the north-east coast. There are, for instance, no shoals of young wrasses (chiefly Jago's goldsinny) gliding amongst the seaweeds, or swarms of grey mullets, as in the tide-runs of the sandy flats and inland seas of the western shores—no large rock-fishes (Ballan wrasses) hiding like dark shadows under the tangles, groups of black gobies between tide-marks, or of young congers breaking the border of the flowing tide into a seething expanse, as in the quiet bays of the southern parts, just as the "sculls" of glittering pilchards do on the surface of the open water. The curious *Hippocampi*, exquisite red mullets, and the splendid conger-fishing also belong to the latter region, together with the abundance of the smaller sharks in-shore. The sandy western shores of England are also distinguished by the greater variety of large Pleuronectidæ, and the frequent occurrence of red gurnards, angelfishes, and spotted rays. The adjoining bay, moreover, does not present that richness of finny life—from the little bimaculated sucker nestling beside its ova in the hollows of the gigantic tangles to the fine cod and coal-fish of the Zetlandic seas, which also possess the rarer *Chimera*. While it is thus vain to look for the vast variety or the gorgeous colours of the species which a few hours' fishing off the shores of Guernsey brings before the investigator, or for the plenitude of large forms which in the north soon fill the boat to overflowing, yet there is sufficient success to reward exertion, either in deep water or off the sea-margin. Good white and flat fish occur in the bay, the latter especially abounding on the sandy flats off the West Sands, which thus form a rich ground for the trawlers, who are for the most part strangers. The trawl in common use (see accompanying figure, and also view of of Harbour at the end of the Fishes) consists of a beam of

wood about 28 feet long, borne on the top of the bulbous ends of two pear-shaped iron structures. A large bag net is fixed to the apparatus, which is dragged behind the boat by ropes attached to the convex portion of the iron supports. The under surface of the latter is flattened, and the point of the apex (which is posterior) turned upwards—the whole thus forming a kind of subaqueous sledge, which glides over the sand and embraces in its progress every thing loose. Young coal-fishes occur all round the rocks and harbour; occasionally a sea-trout is captured off the former; sand-eels frequent the sand near low water; and the salmon-nets are often very productive. In the rock-pools swim hundreds of little two-spotted gobies, swarms of the beautifully coloured young of the lumpsucker, and strings of young sand-eels sport in the sunshine amongst the fringes of seaweeds like flashes of silvery light—affording with other littoral forms, such as shannies and blennies, ample food for the aquatic birds that frequent the beach.

The rarer forms include the lancelet, garfish, doree, opah, oar-fish, and bonito.

z

In the following list the arrangement adopted is that of Dr. A. Günther in his valuable and laborious 'Catalogue of Fishes in the British Museum ' and in subsequent papers.

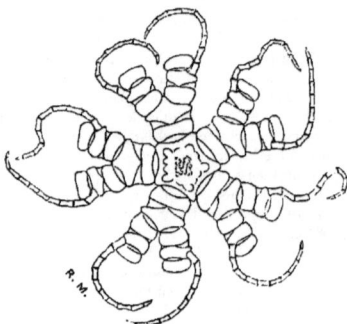

Young *Antedon* (one fifth of an inch), from Lochmaddy.

Subclass I. LEPTOCARDII.

Fam. Cirrostomi.

Genus BRANCHIOSTOMA, Costa.

Branchiostoma lanceolatum, Pall. ; Günther, Catalogue of
the Fishes in the British Museum, vol. viii. p. 513.
Rare. Two specimens occurred in the stomach of a cod.

Subclass II. CYCLOSTOMATA.

Fam. Petromyzontidæ.

Genus PETROMYZON, Artedi.

Petromyzon marinus, L.; Gthr. *op. cit.* viii. p. 501.
Not uncommon. One was captured by attaching itself to
a boat.

Fam. Myxinidæ.

Genus MYXINE, L.

Myxine glutinosa, L.; Gthr. *op. cit.* viii. p. 510.
Occasionally on the cod.

Subclass III. TELEOSTEI.

Order I. ACANTHOPTERYGII.

Fam. Gasterosteidæ.

Genus GASTEROSTEUS, Artedi.

Gasterosteus aculeatus, Albert. Mag. ; Gthr. *op. cit.* i. p. 2.
Frequent on the West Sands after storms.

Gasterosteus spinachia, L. ; Gthr. *op. cit.* i. p. 7.
Abundant in the rock-pools.

z 2

Fam. **Sparidæ**.

Genus PAGELLUS, Cuv. & Val.

Pagellus centrodontus, De la Roche; Gthr. *op. cit.* i. p. 476.
Not uncommon in the bay.

Fam. **Triglidæ**.

Group *COTTINA*.

Genus COTTUS, Artedi.

Cottus scorpius, Bloch; Gthr. *op. cit.* ii. p. 159.
Common in the rock-pools. Small sucking-fishes, shrimps,
Terebellæ, and fragments of green algæ occur in the stomach
of this form.

Cottus bubalis, Euphrasen; Gthr. *op. cit.* ii. p. 164.
Frequent in the rock-pools. Distomes are common in this
species (Plate VIII. figs. 1 & 2). .

Genus TRIGLA, Artedi.

Trigla pini, Bloch; Gthr. *op. cit.* ii. p. 199.
Occasionally procured in the bay.

Tigla hirundo, Bl.; Gthr. *op. cit.* ii. p. 202.
Rare. A single specimen occurs in the University Museum.

Trigla gurnardus, L.; Gthr. *op. cit.* ii. p. 205.
Abundant at all seasons.

Group *CATAPHRACTI*.

Genus AGONUS, Bl.

Agonus cataphractus, L.; Gthr. *op. cit.* ii. p. 211.
Fine specimens are common on the West Sands after
storms.

Fam. **Trachinidæ.**

Group *TRACHININA.*

Genus TRACHINUS (Artedi), Cuv.

Trachinus draco, L.; Gthr. *op. cit.* ii. p. 233.
Frequent on the West Sands after storms.

Trachinus vipera, Cuv. & Val.; Gthr. *op. cit.* ii. p. 236.
Not uncommon in the same locality, and brought in by the fishermen.

Fam. **Scombridæ.**

Group *SCOMBRINA.*

Genus SCOMBER, Artedi.

Scomber scomber, L.; Gthr. *op. cit.* ii. p. 357.
Common.

Genus THYNNUS, Cuv. & Val.

Thynnus pelamys, L.; Gthr. *op. cit.* ii. p. 364.
A fine specimen, about 3 feet long, was procured by Dr. Moir, of St. Andrews, from a salmon-net near the mouth of the Kenley Burn in July 1873, and described by Mr. R. Walker in the 'Scottish Naturalist' for January 1874.

Group *CYTTINA.*

Genus ZEUS (Artedi), Cuv.

Zeus faber, L.; Gthr. *op. cit.* ii. p. 393.
Rather rare.

Group *CORYPHÆNINA.*

Genus BRAMA (Schneid.), Risso.

Brama Raii, Bl.; Gthr. *op. cit.* ii. p. 408.
A specimen occurs in the University Museum.

Genus LAMPRIS, Retzius.

Lampris luna, Retzius ; Gthr. *op. cit.* ii. p. 416.
Rare. A single specimen from the bay exists in the University Museum.

Fam. Carangidæ.

Group CARANGINA.

Genus TRACHURUS, Cuv. & Val.

Trachurus trachurus, L. ; Gthr. *op. cit.* ii. p. 419.
Not uncommon.

Fam. Gobiidæ.

Group GOBIINA.

Genus GOBIUS, Artedi.

Gobius niger, L. ; Gthr. *op. cit.* iii. p. 11.
Mr. Robert Walker states that he has found this species. It has not occurred in my collection.

Gobius Ruthensparri, Euphrasen ; Gthr. *op. cit.* iii. p. 76.
Common in the rock-pools and in the stomachs of cod and haddock.

Group CALLIONYMINA.

Genus CALLIONYMUS, L.

Callionymus lyra, L. ; Gthr. *op. cit.* iii. p. 139.
Common in deep water, and in the stomach of the cod.

Fam. Discoboli.

Group CYCLOPTERINA.

Genus CYCLOPTERUS, Artedi.

Cyclopterus lumpus, L. ; Gthr. *op. cit.* iii. p. 155.
Frequent on the West Sands after storms, and occasionally in the stomach of the cod. The young abound in the rock-pools in autumn.

Group *Liparidina.*

Genus LIPARIS, Artedi.

Liparis vulgaris, Flem.; Gthr. *op. cit.* iii. p. 159.
Occasionally.

Liparis Montagui, Donov.; Gthr. *op. cit.* iii. p. 161.
Abundant in rock-pools, in the laminarian region, and in deeper water, as well as in the stomachs of cod and haddock.

Fam. Pediculati.

Genus LOPHIUS, Artedi.

Lophius piscatorius, L.; Gthr. *op. cit.* iii. p. 179.
Common off the West Sands, and frequently captured in the salmon-nets. One of the specimens had acute pericarditis.

Fam. Blenniidæ.

Genus ANARRHICHAS, Artedi.

Anarrhichas lupus, L.; Gthr. *op. cit.* iii. p. 208.
Frequent in deep water. The stomach of this form contains fragments of *Echinus esculentus, Buccinum undatum, Trochi, Nassa incrassata, Natica, Mya,* starfishes, *Stenorhynchus rostratus,* and *Galathea.*

Genus BLENNIUS, Artedi.

Blennius pholis, L.; Gthr. *op. cit.* iii. p. 226.
[Plate VI. fig. 4.]
Abundant between tide-marks in moist crevices and rock-pools. Feeds on *Baluni,* small littoral shells, and sessile-eyed crustaceans.

Genus BLENNIOPS, Nilss.

Blenniops Ascanii, Walbaum; Gthr. *op. cit.* iii. p. 284.
Not uncommon in deep water, and occasionally in the stomach of the cod.

Genus CENTRONOTUS, Bl.

Centronotus gunellus, L.; Gthr. *op. cit.* iii. p. 285.

Abundant between tide-marks and on the West Sands after storms. The food of this form includes *Hippolyte*, sessile-eyed Crustacea, annelids, starfishes, and small Mollusca (*Rissoa, Skenea*, &c.).

Genus ZOARCES, Cuv.

Zoarces viviparus, L.; Gthr. *op. cit.* iii. p. 295.

Not uncommon between tide-marks and on the West Sands after storms. Sessile-eyed Crustacea and small starfishes occur in its stomach; but in confinement it swallows its fellows.

Fam. Trachypteridæ.

Genus REGALECUS, Brünn.

Regalecus Banksii, Cuv. & Val. (?); Gthr. *op. cit.* iii. p. 309.

An imperfect specimen, 7 feet 2 inches long, occurred amongst the West Rocks, and was described by Mr. R. Walker*.

Fam. Atherinidæ.

Group ATHERININA.

Genus ATHERINA, Artedi.

Atherina presbyter, Cuv.; Gthr. *op. cit.* iii. p. 392.

A specimen in the University Museum. Rare.

Fam. Mugilidæ.

Genus MUGIL, Artedi.

Mugil capito, Cuv.; Gthr. *op. cit.* iii. p. 439.

Not uncommon in the bay.

---

* Ann. & Mag. Nat. Hist. July 1862.

Fam. **Gobiesocidæ.**

Genus LEPADOGASTER, Gouan.

Lepadogaster bimaculatus, Flem.; Gthr. *op. cit.* iii. p. 514.
Occasionally in the laminarian region, and in the stomachs
of the cod and haddock.

Order II. ACANTHOPTERYGII PHARYNGOGNATHI.

Fam. **Labridæ.**

Group *LABRINA.*

Genus LABRUS (Artedi), Cuv.

Labrus maculatus, Bl.; Gthr. *op. cit.* iv. p. 70.
Occasionally brought from deep water. Many young examples occur in the rock-pools in autumn.

Genus CRENILABRUS, Cuv.

Crenilabrus melops, L.; Gthr. *op. cit.* iv. p. 80.
Not common.

Order III. ANACANTHINI.

Suborder **ANACANTHINI GADOIDEI.**

Fam. **Gadidæ.**

Genus GADUS, Artedi.

Gadus morrhua, L.; Gthr. *op. cit.* iv. p. 328.
Common. It is hard to find an inhabitant of the sea that
is not swallowed by this fish.

2 A

Gadus æglefinus, L. ; Gthr. *op. cit.* iv. p. 332.
Common.

Gadus merlangus, L. ; Gthr. *op. cit.* iv. p. 334.
Frequent.

Gadus minutus, L. ; Gthr. *op. cit.* iv. p. 335.
Common.

Gadus luscus, L. ; Gthr. *op. cit.* iv. p. 335.
Not uncommon.

Gadus pollachius, L. ; Gthr. *op. cit.* iv. p. 338.
Occasionally from deep water.

Gadus virens, L. ; Gthr. *op. cit.* iv. p. 339.
Abundant.

Genus MOLVA, Nilss.

Molva vulgaris, Flem. ; Gthr. *op. cit.* iv. p. 361.
Common.

Genus COUCHIA, Thomps.

Couchia argentata, Reinh. ; Gthr. *op. cit.* iv. p. 363.
A single specimen in the stomach of a cod.

Genus MOTELLA, Cuv.

Motella mustela, L. ; Gthr. *op. cit.* iv. p. 364.
Common in rock-pools. Feeds often on sessile-eyed Crustacea.

Motella cimbria, L. ; Gthr. *op. cit.* iv. p. 367.
In a rock-pool at West Rocks. Rare.

Genus RANICEPS, Cuv.

Raniceps trifurcus, Walb.; Gthr. *op. cit.* iv. p. 367.

Not uncommon. The specimens have chiefly been procured from the West Sands after storms.

Genus BROSMIUS, Cuv.

Brosmius brosme, O. F. Müller; Gthr. *op. cit.* iv. p. 369.

Not common.

Fam. **Ophidiidæ.**

Group *AMMODYTINA.*

Genus AMMODYTES, Artedi.

Ammodytes lanceolatus, Lesauvage; Gthr. *op. cit.* iv. p. 384.

Frequent in the débris of storms on the sands, as well as in the latter near low-water mark. Bands of young occur in the tidal pools in May.

Ammodytes tobianus, L.; Gthr. *op. cit.* iv. p. 385.

Occasionally with the former.

Suborder **ANACANTHINI PLEURONECTOIDEI.**

Fam. **Pleuronectidæ.**

Genus HIPPOGLOSSUS, Cuv.

Hippoglossus vulgaris, Flem.; Gthr. *op. cit.* iv. p. 403.

Not rare. The parasitic *Epibdella hippoglossi* is often seen. On the *Caligus* of this fish, *Udonella caligorum* also is common.

Genus RHOMBUS, Klein.

Rhombus maximus, Will.; Gthr. *op. cit.* iv. p. 407.

Common. In the abnormal examples, which swim on their edges, both dextral and sinistral surfaces are coloured, and each has an eye (Plate VI. figs. 5 & 6).

2 A 2

Rhombus lævis, Rondel.; Gthr. *op. cit.* iv. p. 410.
Abundant.

Rhombus megastoma, Donov.; Gthr. *op. cit.* iv. p. 411.
Occasionally.

Rhombus punctatus, Bl.; Gthr. *op. cit.* iv. p. 413.
Not common.

Genus PLEURONECTES (Artedi), Cuv.

Pleuronectes platessa, L.; Gthr. *op. cit.* iv. p. 440.
Abundant. The gulls are fond of emptying the stomachs
of those thrown on the sands after storms.

Pleuronectes limanda, L.; Gthr. *op. cit.* iv. p. 446.
Not uncommon.

Pleuronectes microcephalus, Donov.; Gthr. *op. cit.* iv. p. 447.
Common.

Pleuronectes flesus, L.; Gthr. *op. cit.* iv. p. 450.
Abundant.

Genus SOLEA (Lacép., sp.), Cuv.

Solea vulgaris, Quensel; Gthr. *op. cit.* iv. p. 463.
Frequent.

Solea minuta, Rondel.; Gthr. *op. cit.* iv. p. 470.
Common.

Order IV. Physostomi.

Fam. Salmonidæ.

Genus SALMO, Artedi.

Salmo salar, L.; Gthr. *op. cit.* vi. p. 11.
Many are caught in the stake-nets on the West Sands, and also off the East Rocks.

Salmo trutta, L.; Gthr. *op. cit.* vi. p. 22.
Common.

Genus OSMERUS (Artedi), Cuv.

Osmerus eperlanus, L.; Gthr. *op. cit.* vi. p. 166.
Not uncommon. Large numbers are also caught in the Tay.

Fam. Scombresocidæ.

Genus BELONE, Cuv.

Belone vulgaris, Flem.; Gthr. *op. cit.* vi. p. 254.
Occasionally thrown on the West Sands after storms.

Genus SCOMBRESOX, Lacép.

Scombresox saurus, Walbaum; Gthr. *op. cit.* vi. p. 257.
Not uncommon.

Fam. Clupeidæ.

Group CLUPEINA.

Genus CLUPEA (Artedi), Cuv.

Clupea harengus, L.; Gthr. *op. cit.* vii. p. 415.
Abundant.

Clupea sprattus, L.; Gthr. *op. cit.* vii. p. 419.
Common.

Clupea alosa, L.; Gthr. *op. cit.* vii. p. 433.

Not uncommon. In the stomach of a large specimen caught in the Tay were *Confervæ*, Desmids, and a quantity of vegetable débris.

Fam. **Murænidæ**.

Subfamily *MURÆNIDÆ PLATYSCHISTÆ*.

Group *ANGUILLINA*.

Genus ANGUILLA (Thunberg), Cuv. Règne Anim.

Anguilla vulgaris, Turt.; Gthr. *op. cit.* viii. p. 28.

Abundant in the streams joining the sea, and occasionally between tide-marks; but the latter occurrence is rare in contrast with the western and southern shores.

Anguilla latirostris, Risso; Gthr. *op. cit.* viii. p. 32.

Not uncommon in similar localities. A curious instance of the fatal effects of the voracity of this fish was found in the Swilken burn. A specimen about 20 inches long had seized the head of another not much shorter than itself, and attempted to swallow it. In its struggles the victim's tail also entered the mouth of the larger, and passed out at the left gill-slit, so that several inches were free (Plate VIII. fig. 11). The body of the victim thus formed a large loop which could not be swallowed, while the impaction of the head and tail, and the direction of the teeth of the large specimen, prevented the ejection of the prey. The marauder became exhausted, and was captured at the edge of the stream a the sands.

Genus CONGER, Cuv.

Conger vulgaris, Cuv.; Gthr. *op. cit.* viii. p. 38.

Abundant in deep water.

Order V. LOPHOBRANCHII.

Fam. Syngnathidæ.

Genus SYNGNATHUS, Artedi.

Syngnathus acus, L.; Gthr. *op. cit.* viii. p. 157.
Occasionally seen on the beach after storms.

Genus NEROPHIS (Rafinesque), Kaup.

Nerophis æquoreus, L.; Gthr. *op. cit.* viii. p. 191.
Not uncommon on the West Sands after storms.

Nerophis lumbriciformis, Willughby; Gthr. *op. cit.* viii. p. 193.
Often caught by the hand-net amongst the seaweeds border-
ing pools and rocks as the tide enters; on the West Sands
after storms.

Order VI. PLECTOGNATHI.

Fam. Gymnodontes.

Group MOLINA.

Genus ORTHAGORISCUS, Bl.

Orthagoriscus mola, Bl.; Gthr. *op. cit.* viii. p. 317.
Occasionally caught in the bay. A fine specimen occurred
in October 1862, measuring 4 feet 8 inches from the tip of
the dorsal to the tip of the anal fin, and 3 feet 4 inches from
the snout to the posterior margin of the body. Externally
there were several specimens of *Tristoma coccineum*; and
two wounds existed in the caudal region (from which it was
stated the fishermen pulled two animals like leeches, probably
specimens of *Pontobdella*). Numerous examples of *Gymno-
rhynchus horridus* were found in the muscles on dissection.
The intestine measured 10 feet 6 inches; and the liver weighed
3½ pounds.

Subclass IV. PALÆICHTHYES.

Order GANOIDEI.

Suborder **CHONDROSTEI.**

Fam. **Acipenseridæ.**

Genus ACIPENSER, Artedi.

Acipenser sturio, L.; Gthr. *op. cit.* viii. p. 342.
Occasionally caught in the salmon-nets.

Order CHONDROPTERYGII.

Suborder **PLAGIOSTOMATA.**

A. *SELACHOIDEI.*

Fam. **Carchariidæ.**

Group *CARCHARIINA.*

Genus CARCHARIAS, Cuv.

Carcharias glaucus, L.; Gthr. *op. cit.* viii. p. 364.
Not uncommon in the bay.   Captured by the fishermen in
the salmon-nets.

Genus GALEUS, Cuv.

Galeus canis, Rondel.; Gthr. *op. cit.* viii. p. 379.
Frequently caught in the bay.

Group *MUSTELINA.*

Genus MUSTELUS, Cuv.

Mustelus vulgaris, Müll. & Henle; Gthr. *op. cit.* viii. p. 386.
Not uncommon.

Genus AMPHICTEIS (Gr.), Mgrn.

Amphicteis Gunneri, Sars; Mgrn. *op. cit.* p. 105.

Not uncommon in the stomachs of haddock.

Genus MELINNA, Mgrn.

Melinna cristata, Sars; Mgrn. *op. cit.* p. 106.

Frequent in the stomachs of cod.

Fam. 34. Terebellidæ.

Subfam. 1. *AMPHITRITEA*, Mgrn.

Genus AMPHITRITE, O. F. Müller.

Amphitrite figulus, Dalyell; Mgrn. *op. cit.* p. 107 (as *A. Johnstoni*).

Not uncommon between tide-marks, and ranging to deep water.

Genus LANICE, Mgrn.

Lanice conchilega, Pallas; Mgrn. *op. cit.* p. 108.

Abundant between tide-marks and off the West Sands, and multitudes are thrown on the beach after storms. A common food of many fishes.

Genus NICOLEA, Mgrn.

Nicolea zostericola, Œrst. & Gr.; Mgrn. *op. cit.* p. 109.

Common between tide-marks amongst tangle-roots, and ranging to deep water.

Genus THELEPUS, Leuckart.

Thelepus circinatus, Fab.; Mgrn. *op. cit.* p. 110.

Frequent in the laminarian and coralline regions, in the stomachs of various fishes, and on the West Sands after storms.

S

Subfam. 2. *Polycirridea*, Mgrn.

Genus POLYCIRRUS, Grube.

Polycirrus (Ereutho) Smitti, Mgrn. *op. cit.* p. 111.
Not uncommon between tide-marks.

Subfam. 5. *Canephoridea*, Mgrn.

Genus TEREBELLIDES, Sars.

Terebellides Strœmii, Sars ; Mgrn. *op. cit.* p. 112.
Large specimens occur in the stomachs of cod and haddock.

Fam. 35. Sabellidæ.

Genus SABELLA, L.

Sabella pavonia, Sav. ; Mgrn. *op. cit.* p. 112.
Abundant in the coralline ground, on the West Sands after
storms, and in the stomach of the cod.

Sabella (Branchiomma, Kölliker) *vesiculosa*, Mont. ; Johnst.
Cat. Brit. Mus. p. 259.
Frequently thrown on the West Sands after storms.

Sabella viridis, M.-Edwards, Règ. An. Illust. pl. 1 *e*
(*fide* De Quatref.).
Amongst mud in the interstices of *Filigrana implexa* from
the coralline region.

Genus DASYCHONE, Sars.

Dasychone Dalyelli, Kölliker ; Mgrn. *op. cit.* p. 115.
Occasionally from the coralline ground in the débris of
fishing-boats. .

Genus AMPHICORA, Ehrenberg.

Amphicora Fabricia, O. F. Müller; Mgrn. *op. cit.* p. 117.
Abundant under stones on muddy ground between tide-marks and amongst tangle-roots.

Fam. 36. Serpulidæ.

Genus PROTULA, Risso.

Protula tubularia, Mont. (=*protensa*, Johnst.); Johnst. Cat. Brit. Mus. p. 264.
Occasionally in deep water.

Genus FILIGRANA, Oken.

Filigrana implexa, Berkeley; Mgrn. *op. cit.* p. 119.
Fine masses are common in the coralline region.

Genus HYDROIDES, Gunner.

Hydroides norvegica, Gunner; Mgrn. *op. cit.* p. 120.
Abundant in deep water, attached to shells, stones, &c.

Genus SERPULA, L.

Serpula vermicularis, L.; Mgrn. *op. cit.* p. 120.
Common in deep water.

Genus POMATOCERUS, Phil.

Pomatocerus triqueter, L.; Mgrn. *op. cit.* p. 121.
Very common from the littoral to the coralline region.

s 2

Genus SPIRORBIS, Daud.

Spirorbis borealis, Daud.; Mgrn. *op. cit.* p. 122.
Abundant on seaweeds and stones between tide-marks.

Spirorbis lucidus, Mont.; Mgrn. *op. cit.* p. 123.
Common on zoophytes from deep water.

Series II. ARTHROPODA.

Class CRUSTACEA.

The sessile-eyed Crustacea of St. Andrews are tolerably numerous both in species and individuals. Between tide-marks the most conspicuous (as usual) are the swarms of *Talitrus locusta* which speedily reduce dead fish and other animals to skeletons at high-water mark and considerably beyond it, and the multitudes of *Gammarus locusta* and *Amphithoë podoceroides* under stones amongst the rocks. The *Podoceridæ, Pherusa bicuspis, Calliopius grandoculis,* and *Caprella tuberculata* are plentiful in the rock-pools, and *Corophium grossipes* in the brackish pools near the estuary of the Eden. *Janira maculosa* abounds both in the tidal region and in deep water, while *Jæra Nordmanni* occurs in numbers under stones near high-water mark. In the laminarian region one of the most abundant, perhaps, is *Atylus Swammerdami*, which congregates in swarms on the loose seaweeds. *Siphonœcetus typicus* is common amongst shell-gravel, and *Eurydice pulchra* on the surface of the sea as well as in rock-pools in autumn. Many of the rarer forms occur in the deeper water in considerable numbers; but the distribution of the group in British seas is still involved in considerable obscurity; and at present it will suffice to observe that two of the most plentiful in this region are *Ampelisca Belliana*, Bate, and the new *Calliopius bidentatus*, Norman. The former is likewise common on the beach after storms and in the stomachs of fishes; and the latter ranges to the laminarian zone.

Compared with the Zetlandic area, the absence at St. Andrews of such forms as *Acanthonotus Owenii, Dexamine vedlovensis, Cymodocea truncata,* and *Sphæroma Prideauxianum* in the laminarian region strikes even a superficial observer of the group; while the large number of rare and new species which were met with during the frequent dredgings of Dr. Gwyn Jeffreys and the Rev. A. M. Norman still further heightens the contrast. The southern region, again, is boldly separated by the presence in considerable numbers of *Cymodo-cea truncata* and *Sphœroma Prideauxianum* in the fissures of rocks between tide-marks, and *Dynamene* in rock-pools. The characteristic *Tanais vittatus*, *Paranthura costana*, *Nœsa bidentata*, *Mœra grossimana*, *Chelura terebrans*, *Conilera cylin-drica*, and the large *Cymothoa* parasitic on the fishes at once distinguish the fauna of the Channel Islands from that at St. Andrews. The rarity of *Orchestia littorea* at the latter and its abundance in the tidal region of the Outer Hebrides, and the absence of *Sulcator arenarius* and its frequent occur-rence in the sand of the western shores of England, are also interesting contrasts.

Many of the sessile-eyed Crustacea, such as *Talitrus locusta*, are extremely hardy. *Gammarus locusta* is often found in putrid localities, and it survives almost every other marine form in putrid vessels in confinement. The group as a whole is composed of extremely active animals ; and even the most grotesque, such as *Caprella tuberculata*, are at home in the intricacies of *Ceramium* and other finely branched seaweeds. The boring forms (by jaws) are represented by *Limnoria lignorum* ; but its depredations are comparatively insignificant, probably because little wood is employed within water-mark in the construction of the harbour. The perforations of *Talitrus*, again, abound in the sand, and the looped burrows of *Coro-phium* in the sandy mud of the flats it inhabits. The nest-forming crustaceans are represented by *Amphithoë podoceroides*, *Siphonœcetus typicus*, *Podocerus variegatus*, and *P. falcatus* ; while the young of *Gammarus locusta* are often observed adhering to the abdominal region of the parent.

The Cirripedes occur abundantly between tide-marks, the most conspicuous being *Balanus balanoides*, which covers the bare rocky ridges opposite the Castle and other parts. In deep water the various species are attached to shells, stones, crabs, wood, cork, coal, tests of ascidians, and other structures.

I am indebted to Mr. Spence Bate for the determination of several doubtful forms, and especially to the Rev. A. M. Norman for his courteous assistance in this respect, and in revising the list. Mr. G. S. Brady kindly furnished me with the names of the Ostracoda occurring in shell-débris on the West Sands and other collections.

Order PYCNOGONOIDEA.

Fam. **Pycnogonidæ**, Latreille.

Genus PYCNOGONUM, Brünnich.

Pycnogonum littorale, O. F. Müller.

Abundant under stones between tide-marks.

Genus PHOXICHILIDIUM, M.-Edwards.

Phoxichilidium femoratum, Rathke.

Occasionally under stones in rock-pools, and ranging to deep water.

Besides the foregoing, there are several species (one apparently identical with Mr. Goodsir's *Nymphon Johnstoni*, and another with his *N. spinosum*) not uncommon in the coralline region. Many delicate zoophytes are found on their limbs.

Order CIRRIPEDIA.

Suborder **SUCTORIA**.

Fam. **Peltogastridæ**, Claus.

Genus PELTOGASTER, H. Rathke.

Peltogaster paguri, H. Rathke.

Occasionally on *Pagurus bernhardus*. A more elongated form occurs on *P. cuanensis*.

Genus SACCULINA, Thompson.

Sacculina carcini, Thompson.

[Plate IX. fig. 13.]

Common on the abdomen of *Carcinus mænas*. Another is found on *Portunus holsatus* (Plate IX. figs. 14 & 15).

Suborder **THORACICA**.

Fam. Lepadidæ.

Genus LEPAS, L.

Lepas anatifera, L.; Darwin, Mon. i. p. 73, pl. 1. f. 1.

On the bottoms of ships, and thrown ashore after storms attached to timber.

Genus SCALPELLUM, Leach.

Scalpellum vulgare, Leach; Darw. Mon. i. p. 222, pl. 5. f. 15.

On *Thuiaria thuja* and *Sertularia cupressina* from deep water.

Fam. Balanidæ.

Subfamily *BALANINÆ*.

Genus BALANUS, Lister.

Balanus porcatus, E. da Costa; Darw. Mon. ii. p. 256, pl. 6. f. 4.

Abundant on stones, *Ascidia sordida*, crabs, &c. in deep water, and occasionally between tide-marks.

Balanus crenatus, Bruguière; Darw. Mon. ii. p. 261, pl. 6. f. 6.

Not uncommon on *Hyas araneus*, *Lithodes maia*, and on rocks in the laminarian region.

Balanus balanoides, L.; Darw. Mon. ii. p. 267, pl. 7. f. 2.

Very abundant; coating extensive surfaces of the rocks between tide-marks and in the laminarian region, and adhering to mussels, sticks, posts, &c. Elongated varieties are not uncommon. The exuviæ swarm in the rock-pools and on the surface of the sea in summer.

Balanus Hameri, Ascanius; Darw. Mon. ii. p. 277, pl. 7. f. 5.

Occasionally in deep water; a small thorn-tree (still fresh) was covered with fine examples.

Fam. **Scylliidæ.**

Genus SCYLLIUM, Cuv.

Scyllium canicula, L.; Gthr. *op. cit.* viii. p. 402.
Not uncommon in the bay.

Fam. **Spinacidæ.**

Genus ACANTHIAS, Risso.

Acanthias vulgaris, Risso; Gthr. *op. cit.* viii. p. 418.
Often caught on the deep-sea lines of the fishermen.

Genus LÆMARGUS, Müller & Henle.

Læmargus borealis, Scoresby; Gthr. *op. cit.* viii. p. 426.
Occasionally caught off the bay, near the estuary of the
Forth.

B. *BATOIDEI.*

Fam. **Rajidæ.**

Genus RAJA, Artedi.

Raja clavata, L.; Gthr. *op. cit.* viii. p. 456.
Frequent.

Raja radiata, Donov.; Gthr. *op. cit.* viii. p. 460.
Not uncommon on the sandy flats.

Raja circularis, Couch; Gthr. *op. cit.* viii. p. 462.
Not uncommon.

Raja batis, L.; Gthr. *op. cit.* viii. p. 463.
Common.

2 B

Raja lintea, Fries ; Gthr. *op. cit.* viii. p. 466.

This and the following are entered on the authority of Mr. R. Walker, Librarian of the University.

Raja fullonica, L. ; Gthr. *op. cit.* viii. p. 467.

Occasionally seen.

PLATE II.

Fig. 1. *Edwardsia callimorpha*, Gosse, in its investing tube.   Enlarged.

Fig. 2. The same, in a slightly contracted condition.

Fig. 3. *Edwardsia Allmanni*, M'I., in its investing tube.   Considerably enlarged.

Fig. 4. *Edwardsia Goodsiri*, M'I.   Enlarged.

Fig. 5. *Peachia hastata*, Gosse, in the semicontracted state.   Slightly enlarged.

Fig. 6. The same species in expansion.   Similarly enlarged.

Fig. 7. The same (a contracted specimen) with adherent particles forming a kind of tube.   Enlarged.

Fig. 8. A young example of *Actinoloba dianthus*, attached to the side of a glass vessel.

Fig. 9. *Clavelina lepadiformis*, O. F. Müller.   Enlarged.

Fig. 10. *Pelonaia corrugata*, Forbes & Goodsir.   Slightly enlarged.

Fig. 11. *Eolis Adelaidæ*, Thompson.   Enlarged.

Fig. 12. *Eolis Farrani*, A. & H., purple variety.

Fig. 13. An example of the same species more richly tinted.

Fig. 14. *Doto coronata*, Gmelin, with abnormal left dorsal tentacle and sheath.

Fig. 15. *Tritonia plebeia*, Johnston, bifid posteriorly.

Fig. 16. *Doris Johnstoni*, A. & H.   Slightly enlarged.

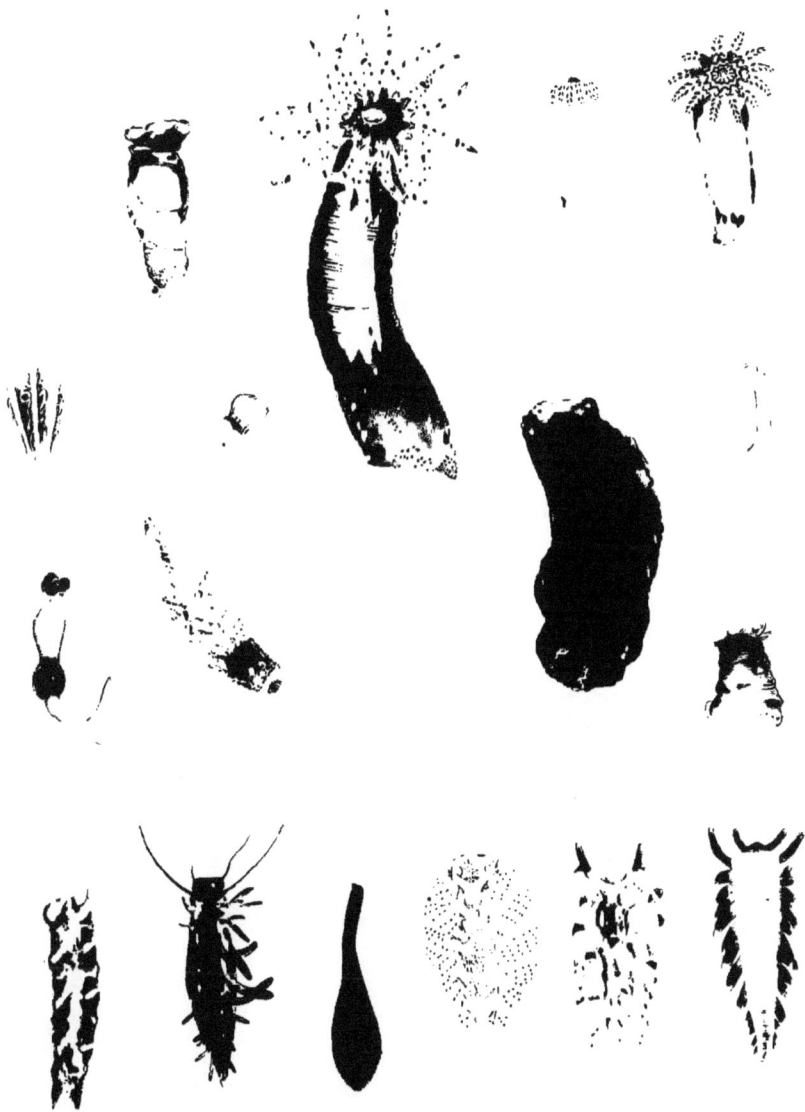

PLATE III.

Fig. 1. *Aplysia punctata,* Cuv., in the contracted state. Somewhat reduced. The great muscular lobe (epipodium) of the right side has been folded downwards, so as to expose the shell (on the left in the figure). The purplish dye exuded by the animal surrounds it.

Fig. 2. *Lamellaria perspicua,* L. Slightly enlarged.

Fig. 3. Another specimen. About natural size.

Fig. 4. The same, slightly contracted. Viewed from the under surface.

Fig. 5. A pale specimen of the same species. Slightly enlarged.

Fig. 6. View of the under surface of the same.

Fig. 7. Anterior view of another small specimen, showing the foot and cloak.

Figs. 8, 9 & 10. Figures showing the great variability of colour in this species.

Fig. 11. Lateral view of *Lucernaria auricula,* O. Fabr., attached to a piece of *Fucus serratus.* Enlarged under a lens. The pale streak on the pedicle indicates one of the longitudinal muscles.

Fig. 12. View of the same expanded, and seen from above. Similarly enlarged. Running inwards from each bundle of tentacles are the orange reproductive organs. One of the border papillæ is double; and various urticating organs are scattered over the bell.

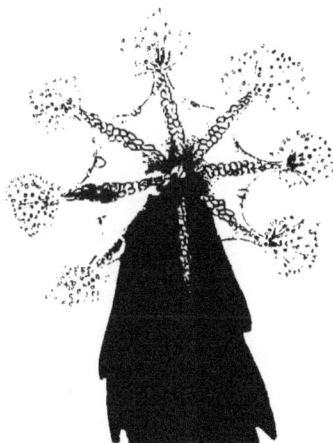

PLATE IV.

Fig. 1. *Echiurus vulgaris,* Sav. About natural size.

Fig. 2. *Priapulus caudatus,* Lam. The specimen is somewhat contracted, so as to give prominence to the transverse rugæ.

Fig. 3. A very large specimen, apparently of *Amphiporus pulcher.* Enlarged.

Fig. 4. *Synapta inhærens,* O. F. Müller. Somewhat enlarged.

Fig. 5. *Cucumaria lactea,* Forbes & Goodsir, attached to a glass vessel. Enlarged.

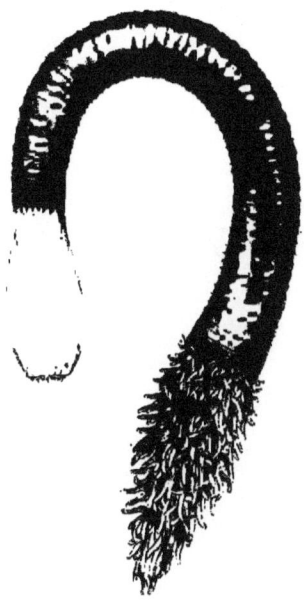

PLATE V.

Fig. 1. *Pontobdella muricata*, L. Slightly enlarged.

Fig. 2. *Piscicola geometra*, L. Enlarged.

Figs. 3, 4, & 5. *Pontobdella littoralis*, Johnst., various positions, on *Cottus bubalis*.

Fig. 6. Sucker of the same. All enlarged.

Fig. 7. First pair of foot-jaws in a brilliantly-coloured male of *Carcinus mœnas*, L.

Fig. 8. Under surface of the carapace of the same male.

Fig. 9. *Frontonum* (?) from deep water.

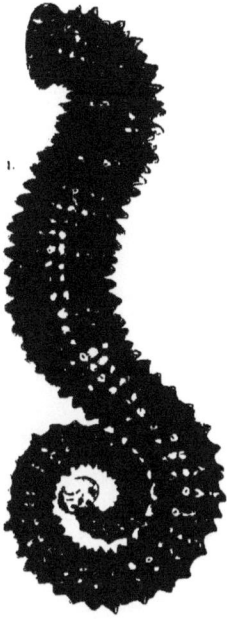

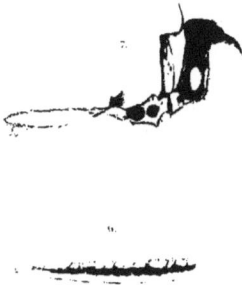

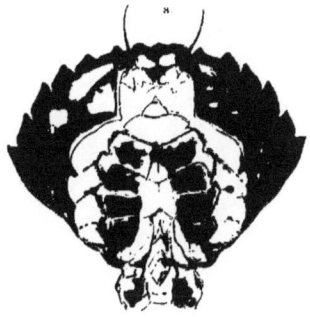

PLATE VI.

Fig. 1. Monstrosity of *Asterias rubens* (dried), in which one arm has become bifid. Natural size.

Figs. 2 & 3. Other examples of the same species, in which peculiar appearances arise from the reproduction of arms.

Fig. 4. Shanny (*Blennius pholis*) emerging from a tuft of *Halidrys siliquosa*.

Figs. 5 & 6. Right and left (usually upper and lower) sides of an abnormal Turbot. Both sides are coloured; and the right eye has been arrested in its progress towards the left or upper surface. From a spirit preparation.

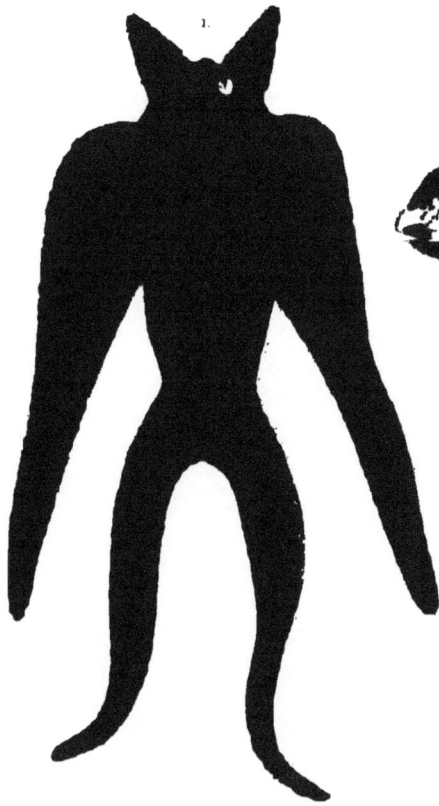

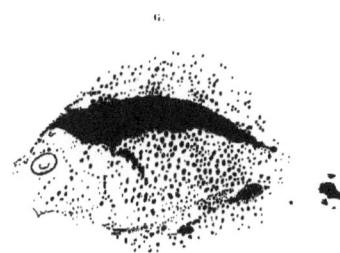

PLATE VII.

Fig. 1. Oral disk of *Edwardsia Allmanni*, M'I., completely contracted. Enlarged.

Fig. 2. The same, with the tentacles partially expanded.

Fig. 3. Anterior portion of the sheath and pouting rim of the mouth. Viewed laterally.

Fig. 4. A single tentacle. Enlarged.

Fig. 5. View of the posterior part of *Edwardsia Goodsiri*, M'I., showing the "blown-bladder" aspect.

Fig. 6. The oral disk with expanded tentacles.

Fig. 7. A single tentacle. Enlarged.

Fig. 8. An entire colony of *Alcyonium digitatum*, L., with the polyps in varying degrees of expansion.

Fig. 9. External view of a polyp on a level with its cell.

Fig. 10. The same, more expanded.

Fig. 11. A polyp half-expanded.

Fig. 12. A polyp of the same, much attenuated, showing the minute spicula and the somewhat fusiform aspect of the expanded arms.

Fig. 13. *Doris repanda*, A. & H., in process of spawning : *a*, the dimple of the foot. The arrow marks the direction of its turning. Seen through a glass vessel.

Fig. 14. The same, in another position, with the cloak much folded.

Fig. 15. Spawn of the foregoing. The single interrupted chain of ova appended thereto is exceptional.

Fig. 16. A scolex of *Tetrarhynchus* from the stomach of *Cancer pagurus*, encysted. Probably from the débris of fishes eaten by the crab. × about 60 diameters.

Fig. 17. The same, separated from the sheath and somewhat compressed. × 65 diam.

Fig. 18. *Echinorhynchus strumosus*, Rud., from the intestine of a common seal (*Phoca vitulina*) caught at St. Andrews. × 60 diam.

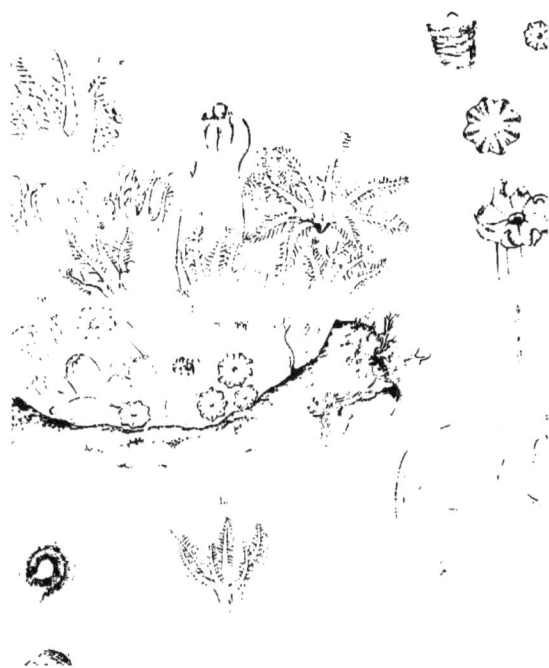

PLATE VIII.

Fig. 1. *Distoma* from muscles of *Cottus bubalis*. × 60 diameters.

Fig. 2. Appearance of the posterior end of the *Distoma* after pressure. Similarly magnified.

Fig. 3. *Mesostomum bifidum.* n. sp. × 60 diam.

Fig. 4. Penis of the foregoing. × 350 diam.

Fig. 5. Rod-shaped bodies and granules from the testes of the specimen. × 350 diam.

Fig. 6. Fully developed spermatozoa of the same animal.

Fig. 7. *Vortex capitata.* (Erst. × about 200 diam.   *a*, muscular proboscidian organ, not ciliated ; *a'*, secondary and probably exsertile muscular organ ; *j*, œsophageal division ; *j'*, central digestive chamber ; *j''*, posterior digestive chamber ; *z*, anus ; *m*, cilia of the organ homologous with the cephalic sac of the Nemerteans ; *d*, clear papilla with active though short cilia (in some views a duct-like structure passes inwards from *d*) *h*, clear globule capable of contraction under pressure ; *i*, radiate appearance at genital sphincter (?) ; *k*, sperm-sacs ; *o*, mouth.

Fig. 8. Another specimen, showing ova. Similarly magnified.   *ac*, ova.

Fig. 9. Cells of the alimentary region. × 350 diam.

Fig. 10. Spermatozoa of *Vortex capitata*. × 700 diam.

Fig. 11. View of an Eel (which was choked while endeavouring to swallow a fellow eel).

PLATE IX.

Fig. 1. *Ascidia sordida*, attached to the upper end of the tube of *Pectinaria belgica*. Very slightly enlarged.

Fig. 2. The same, from the other side.

Fig. 3. *Parascidia Flemingii*, Alder, attached to a rootlet of tangle.

Fig. 4. Egg-case of *Nassa incrassata*, Ström. Enlarged.

Fig. 5. Tentacles of *Cucumaria lactea*. Enlarged. The oral orifice occupies the centre.

Figs. 6 & 7. Views of the oral region and tentacles of *Synapta inhaerens* in different degrees of contraction.

Fig. 8. Anchor and plate of the foregoing. × 280 diameters.

Fig. 9. Section of the shell of *Cancer pagurus*, showing the internal processes of the hairs. The preparation was mounted by the late Dr. Fraser Thomson.

Fig. 10. Young of common Mussel attached by its byssus to the carapace of a Shore-Crab.

Fig. 11. Young Mussels in the eye-cavities of the same crustacean, causing the peduncles to project, and partially blinding the Crab.

Fig. 12. Defect of the carapace in the Shore-Crab, exposing the branchial chamber exterior to the first pair of foot-jaws.

Fig. 13. Abdomen and part of the cephalothorax of a female Shore-Crab, with a parasitic *Sacculina carcini* and other abnormalities.

Fig. 14. *Sacculina* parasitic on a male *Portunus holsatus*.

Fig. 15. Another view of the same, with the tail reflected.

www.ingramcontent.com/pod-product-compliance
Lightning Source LLC
Chambersburg PA
CBHW021656210326
41599CB00013B/1444